THE TRADITIONAL CHINESE IRON INDUSTRY AND ITS MODERN FATE

NORDIC INSTITUTE OF ASIAN STUDIES
Recent NIAS Reports

23. MANAGING THE IRRIGATION: THARU FARMERS AND THE IMAGE OF THE COMMON GOOD
    Sven Cederroth

24. THE NEXT LEFT? DEMOCRATISATION AND ATTEMPTS TO RENEW THE RADICAL POLITICAL DEVELOPMENT PROJECT – THE CASE OF KERALA
    Olle Törnquist

25. THE QUEST FOR BALANCE IN A CHANGING LAOS: A POLITICAL ANALYSIS
    Søren Ivarsson, Thommy Svensson and Stein Tønnesson

26. AIDING AFGHANISTAN. THE BACKGROUND AND PROSPECTS FOR RECONSTRUCTION IN A FRAGMENTED SOCIETY
    Asger Christensen

27. MEASURING ATTITUDES IN EAST ASIA: THE CASE OF SOUTH KOREAN DEMOCRATIZATION
    Geir Helgesen and Søren Risbjerg Thomsen

28. VIETNAM OR INDOCHINA? CONTESTING CONCEPTS OF SPACE IN VIETNAMESE NATIONALISM, 1887-1954
    Christopher E. Goscha

29. ISLAM, HUMAN RIGHTS AND CHILD LABOUR IN PAKISTAN
    Alain Lefebvre

30. NATURAL RESOURCES AND COSMOLOGY IN CHANGING KALASHA SOCIETY
    Mytte Fentz

31. MIGRATION IN CHINA
    Børge Bakken

32. TRADITIONAL CHINESE IRON INDUSTRY AND ITS MODERN FATE
    Donald B. Wagner

The Traditional Chinese Iron Industry and Its Modern Fate

Donald B. Wagner

with a Foreword by Peter Nolan

Routledge
Taylor & Francis Group

LONDON AND NEW YORK

Nordic Institute of Asian Studies
NIAS Report series, No. 32

First published in 1997
by Curzon Press

Published 2023 by Routledge
4 Park Square, Milton Park, Abingdon, Oxon OX14 4RN
605 Third Avenue, New York, NY 10017

*Routledge is an imprint of the Taylor & Francis Group, an
informa business*

Typesetting by the Nordic Institute of Asian Studies

British Library Catalogue in Publication Data
A CIP catalogue record for this book
is available from the British Library

Library of Congress in Publication Data
A catalog record for this book has been requested

ISBN 13: 978-0-7007-0951-9 (pbk)

In memory of my mother
Edith Winifred Belk Wagner
1917 – 1992

Contents

Figures ... viii

Foreword ... ix

Acknowledgements ... xiii

1. Introduction ... 1

2. The changing economic geography of the traditional Chinese iron industry ... 4
    Western competition ... 5
    Return on investment ... 8
    The effects of decline on technology ... 9
    The Great Leap Forward ... 10

3. Traditional Chinese iron production techniques ... 12

4. Small-scale ironworks of the mountains of Dabieshan ... 15
    The blast furnace ... 16
    The fining hearth ... 21
    The product ... 25
    The iron industry of Dabieshan ... 26

5. Large-scale ironworks in Sichuan ... 29
    The blast furnace ... 31
    The puddling furnace ... 41
    The iron industry of Sichuan ... 43

6. Crucible smelting in Shanxi ... 48
    Crucible smelting ... 48
    The iron industry of Shanxi ... 53

7. Large- and small-scale ironworks in Guangdong ... 58
    Small blast furnaces ... 58
    Large blast furnaces ... 63
    The iron industry of Guangdong ... 71

8. Concluding remarks ... 76

Bibliography ... 80

Index ... 103

Figures

1 Flow diagram of traditional Chinese blast-furnace iron production ... 12
2 Blast furnace in Xinyang, Henan, ca. 1916 ... 16
3 Blast furnace in Xinyang, Henan, ca. 1916 ... 17
4 Blast furnace in Jinzhai, Anhui, ca. 1958 ... 18
5 Plan of a 'smelting house' at East Wind People's Commune in Muzidian, Macheng County, Hubei ... 19
6 Sketch of the Huang Jiguang blast furnace in Macheng, Hubei, as used in the Great Leap Forward period, and associated requisites ... 19
7 Vertical section of the Huang Jiguang blast furnace in Macheng, Hubei, as used in the Great Leap Forward period ... 20
8 Diagram of a modern large-scale coke-fuelled iron blast furnace ... 22
9 Diagram of a fining hearth used in Shangcheng, Henan, in 1958 ... 24
10 Drawing by E. T. Nyström of a pair of fining hearths in southern Henan, ca. 1916 ... 25
11 Sketch and sections of a water-powered blast furnace at Huangnipu in Rongjing County (modern Yingjing), Sichuan, ca. 1877 ... 32
12 Vertical section through a blast furnace in Qijiang County, Sichuan ... 34
13 Photograph of a blast furnace somewhere in Sichuan ... 35
14 Internal form of two traditional blast furnaces in Sichuan in 1940 ... 38
15 Diagram of a puddling hearth in Sichuan ... 40
16 Puddling furnace used in Sichuan in 1958 ... 43
17 Stall furnace for crucible iron smelting in southern Shanxi ... 49
18 Lumps of cast iron and puddled wrought iron in southern Shanxi ... 49
19 Crucible smelting of iron in progress in Gaoping County, Shanxi ... 50
20 Operation of a furnace in Shanxi for converting cast iron to wrought iron ... 52
21–28 First eight of twelve watercolours in an album preserved in the Bibliothèque Nationale, Paris, and painted by a Chinese artist in Guangzhou in the middle of the nineteenth century ... 59–62
29 One possible reconstruction of an early Qing blast furnace investigated at Luxia Village in Luoding County, Guangdong ... 69

Foreword

It is a great pleasure to have been asked to write this foreword to Don Wagner's latest piece of research. Ferrous metals are at the heart of economic development, ancient and modern. These constitute the basic input for a wide range of consumer durables, for machinery, and for infrastructure construction. Don's research over many years has shed great light upon this key sector. His monumental work, *Iron and Steel in Ancient China*,[1] has quickly become the standard source on the topic. His monograph on Dabieshan is a pioneering piece of modern economic history, unusual in its linking of pre- and post-1949 analysis in a single piece of work.[2]

Throughout his research he has insisted that it is necessary to understand the technical basis of the production of ferrous metals in order to analyse the economic history of this subject. It is our good fortune that, like his mentor Joseph Needham, Don combines great scholarship and scientific rigour with exceptional clarity of presentation, so that those like myself, who were not trained as scientists, can nevertheless follow the argument. Don has also insisted on the importance of situating technical change in the wider socio-economic context.

This monograph carefully examines China's modern iron industry in a regional setting. Through an examination of its iron industry, Don's book brings wonderfully alive the regional differences in China's modern economy. In the Introduction he hints at a possible future work in which he may analyse in detail the story of the iron industry in one of the four regions examined in this book, including an analysis of the role of governments at different levels and the role of entrepreneurs. This is greatly to be looked forward to.

◆ ◆ ◆

---

1. Leiden, E. J. Brill, 1993.
2. *Dabieshan: Traditional Chinese iron-production techniques practised in southern Henan in the twentieth century*, Curzon Press, London and Malmö, 1985.

ix

The history of China's ferrous metals sector parallels the history of the whole Chinese economy. Iron was a relatively late introduction to China, with a true Iron Age beginning around the sixth century BC, long after the Iron Age of the West; but the Chinese developed the ability to *cast* iron almost as soon as they knew about it at all. Moreover, *steel* production did not lag behind that of iron. By the sixth century AD China was producing steel by an ancestor of the Siemens–Martin open hearth process. The wide application of iron and steel (as opposed to its use by a small elite) came much earlier in China than in Europe.

There is still great uncertainty about the long-run trend in China's output of ferrous metals from the tenth or eleventh century AD. It is uncertain at what point Europe took over from China as the world's leading region in their production. In Don's judgement China had 'the world's largest and most efficient iron industry' until around 1700, but after that point an 'extraordinary sequence of technical improvements' brought down the price of iron dramatically and was a leading factor in the British Industrial Revolution. The central theme of this book is the way in which the iron industry in different parts of China was affected by this severe challenge from the outside world after the late eighteenth century.

By the 1930s China's steel output was only a small fraction of that of the world's leading steel producers. In 1936 China produced just 0·4 million tonnes, compared to over four million tonnes in Russia, over seven million tonnes in France and 27 million tonnes in Germany. It was at about the same level as a small European economy such as Austria. After 1949, China began its long process of catch-up in ferrous metals, both technically and in terms of total output. By the late 1970s, China ranked fifth in the world in terms of total output.

Under accelerated modernisation from the 1980s, China's rapid growth in industrial output was matched by a comparable growth in iron and steel output. While output in the advanced capitalist economies stagnated, and collapsed in the former USSR, output in China grew apace to meet the booming demand for iron- and steel-using products, from motor vehicles, refrigerators and electric fans, to bridges, ships, and apartment blocks. China's steel output rose from 36 million tonnes in 1980 to 90 million tonnes in 1993.[3]

Alongside the expansion of output went rapid technological upgrading. Huge, completely modern plants were built, such as Baogang in Shanghai and many large older plants, such as Shougang in Beijing, comprehensively modernised in the reform period. An important part of the modernisation of such plants was 'learning by doing' through the purchase, dismantling and

---

3. Ministry of Metallurgical Industry (*Yejin Bu* 冶金部), *Zhongguo gangtie tongji* 中國 鋼鐵統計 (Statistics on China's steel industry), Beijing, Ministry of Metallurgical Industry, 1994, pp. 53–55.

reassembling of modern, second-hand equipment from the stagnating steel industries of Europe and the USA.[4] As this monograph demonstrates for an earlier period, the story of China's recent growth in iron and steel is one in which growth, modernisation and technical progress interact with real historical processes. An analysis of government action at different levels, of regional peculiarities and of the role of industrial entrepreneurs who run China's iron and steel plants, continues to be necessary in order to understand the way in which China's ferrous metals industry is evolving.[5]

By the early 1990s China had surpassed the USA to become the world's second largest producer and was poised to overtake Japan as the leading producer. However, China's per-capita steel output was still only a tiny fraction of that of the advanced economies, standing at around 60 kilograms, compared to over 400 kilograms in the USA, 800 kilograms in Japan and over 1,200 kilograms in Singapore.[6] It is likely that China's economy will continue to grow at a fast rate, and that demand for steel will also grow rapidly. At some point in the early- to mid-twenty-first century it is probable that China's steel output will surpass 150 million tonnes, and may well be much above 200 million tonnes.[7] China will then have truly recaptured the central position in the world's ferrous metals industry that it occupied for such a long period of medieval history.

In China's recent accelerated growth of iron and steel output, local plants of the kind that form the focus of analysis in this monograph continue to make an important contribution to China's ferrous metals output. In 1992 there were over 1700 steel 'plants' in China. Although large 'keypoint' plants dominated, with almost 70 per cent of total output, smaller plants were sharply increasing their contribution. Over 1,600 local plants accounted for around one-third of total output in the early 1990s.[8] The 140-odd keypoint plants averaged around 12,000 workers each and on average produced an annual output of around 375,000 tonnes of steel. Over 1,600 local plants averaged only around 1,000 workers each, and averaged only around 14,000 tonnes each per annum. Some of the local plants were modern 'mini-mills'

---

4. This was the case also for Posco, South Korea's giant steelmaker.
5. See, for example, Peter Nolan, 'From state factory to modern corporation? Shougang Iron and Steel Corporation under economic reform', mimeo, 1996.
6. Ministry of Metallurgical Industry, op. cit., pp. 130–131.
7. Tan Chengdong, 'Optimising the structure is the main target of developing the iron and steel industry in the socialist market system', in Research Centre of Ferrous Metallurgy, Ministry of Metallurgical Industry, ed., Yejin jingjixue 冶金經濟學 (Metallurgical economics), No. 16, 1994, pp. 15–22.
8. Ministry of Metallurgical Industry (Yejin Bu 冶金部), Zhongguo gangtie tongji 中國鋼鐵統計 (Statistics on China's steel industry), Beijing, Ministry of Metallurgical Industry, 1993, section 1.

using sophisticated modern techniques, typically with electric furnaces, and using scrap steel. However, many of them were integrated plants based on older, local, and typically more labour-intensive, techniques of the kind analysed in this monograph. They used local iron ore mined in handicraft mines, often with rough and dangerous conditions of labour. They typically converted local pig iron into steel. Technical and socio-economic analysis of the recent history of this huge sector, comparable to Don Wagner's for an earlier period, hardly exists.

This monograph is an invaluable work of technological and economic history. However, as I hope this Foreword has shown, it has great contemporary relevance. Ferrous metals have been at the centre of world economic evolution, and will continue to be so for a long time to come. This monograph is centrally concerned with the response of China's ferrous metals industry to the challenge of Western producers from the late eighteenth century, who dominated world production and technical change for the next two hundred years. Having been the world leader for well over a thousand years, China ceased to be so for what must now be seen as a relatively brief period of time. By the 1990s China had resumed its central importance in the world's ferrous metals industry; in the future its role is likely to become ever larger, both in its share of world output and in its contribution to world technical progress in ferrous metals.

Peter Nolan
Jesus College, Cambridge

Acknowledgements

Numerous friends have given me useful comments on earlier drafts of this book, including Francesca Bray, Christopher Cullen, Charles Curwen, Peter Golas, J. R. Harris, Graham Hollister-Short, Peter Nolan, and Joanna Waley-Cohen. As always it is necessary to emphasise that I alone am responsible for errors, misunderstandings, and infelicities of expression that remain.

This book will, in revised form, form a chapter in the volume on Ferrous Metallurgy of Joseph Needham's *Science and Civilisation in China*, which will not be ready for publication for some time. I am grateful to the Cambridge University Press and the Publications Board of the Needham Research Institute for permission to publish it separately.

The actual writing has been done under a grant from the Leverhulme Foundation; much preliminary research was done under earlier grants from the Danish Research Council for the Humanities, the Carlsberg Foundation, the University of Copenhagen, and the Julie von Müllen Foundation. During spells of unemployment the dole was supplemented by my dearest friend, Annie Winther.

Map of eastern China, showing all of the modern place names mentioned in the text. The four regions discussed are the provinces of Sichuan, Shanxi, and Guangdong and the Dabieshan mountains, where the provinces of Henan, Hubei, and Anhui meet.

1

Introduction

This little book is a preliminary exploration of a very large subject, the economic history of the Chinese traditional iron industry in the nineteenth and early twentieth centuries. It is especially concerned with the ways in which technological choices interact with other historical factors: the technologies used in any particular time and place are the product of a historical development strongly influenced by geographic and economic factors (among others); and in turn these technological choices can have a strong influence on broader historical developments.

I consider here the traditional iron industries of four regions of China, chosen for the availability of sources and for the variety of technologies which they exhibit. For each place I describe in detail the technologies used and investigate briefly its history in the past few centuries, then consider in more detail the particular question of the effect of modern competition on the industry since the middle of the nineteenth century.

This study began for me with a problem in historical methodology. We have a large number of good sources on Chinese iron production technologies in this period. In particular, there are eyewitness accounts by travellers, including a number of Chinese and Western engineers whose technical expertise allowed them to understand what they were seeing. It has sometimes been believed that these descriptions give us a direct view of technologies of very ancient times. One of the best of our eyewitnesses, Thomas T. Read, who worked in China 1907–10, was also one of the first to realise that this cannot be the whole story. As we learn more about the ancient iron industry it becomes clear that there is no immediate or simple correspondence between its technologies and the technologies of recent centuries. Read presented this problem in an important article entitled 'Chinese iron, a puzzle' (1937), but apparently never found an explanation that really satisfied him.

Historians of technology are today more willing to accept that technological development need not run in a straight line from 'primitive' to 'sophisticated'. While there may (or may not) be a general accretion of knowledge and techniques within the industry, there is no easy way of com-

1

paring the relative 'sophistication' of two different technologies. In particular, the tendency to consider a small-scale labour-intensive technology to be more primitive than one which is large-scale and capital-intensive is not always just, as some of the examples in this book should show. Technologies must be judged in the context in which they are used; in particular, *efficiency* is difficult (perhaps impossible) to define in a way which is meaningful across widely differing economic and geographical contexts.

In the present book I believe I have uncovered one corner of a solution to Read's puzzle. From the eighteenth century onward, and with particular haste from the late nineteenth century, various aspects of contact with the West led to great changes in the economic geography of the Chinese iron industry. I shall place most emphasis here on competition with cheap Western iron, which led to reduced prices and profits, but other aspects were also important. The tea and opium trades provided new opportunities for investment of capital; steamships and railways reshaped marketing patterns; and the Chinese state, occupied with meeting the imperialist challenge, found itself more in need of a strong industry but much less able to intervene successfully in its operation.

The new conditions brought by these changes strongly favoured labour-intensive technologies. In the language of economics, the industry was pushed to the labour-intensive end of its production possibilities curve. In regions where this curve was discontinuous (that is, two or more different technologies were in use), as in Guangdong, the more capital-intensive technologies disappeared entirely.

Further violent changes in the economic geography of Chinese industry were caused by World War I, the Sino–Japanese War, and China's isolation in the first decades after 1949. These events should have made the larger-scale and more capital-intensive of the traditional technologies more attract-ive, but by the twentieth century those technologies were largely forgotten. Occasional attempts to improve the technologies by the application of modern technical expertise appear to have had little lasting effect. The last episode in the modern fate of the traditional iron industry occurred during the Great Leap Forward of 1958–60, when the traditional iron-production technologies played a part in an enormous effort to expand iron and steel production and bring China out of a situation of economic gridlock. That campaign was overall a massive failure, but it had some partial local successes which have generally been overlooked. And it also provided some of the best sources for the study of the traditional techniques, for hundreds of careful technical studies of traditional technologies were published in connection with the campaign.

I have already called this book a preliminary exploration. It is strongest on the technologies, and approaches economic issues largely in terms of the long-term factors of neo classical economic theory. To do proper justice to the subject a great deal more work must be done on the roles of individual actors, especially the entrepreneurs and governments at various levels. Four solid monographs, one for each of the regions considered here, would be needed; if I live long enough I may attempt to write one of them.

I have not skimped on technical detail, but I have done what I can to make it less painful to the reader. I have tried to give explanations which will be adequate for readers who know some chemistry and are accustomed to technical thinking. Others will find, I hope, that they are able to skip over the most technical parts without losing the thread of the argument – but I would also urge them to make a try, for technical thinking is a fundamental aspect of the modern world, and historians and social scientists who attempt to ignore it are in danger of losing a much more important thread.

2

The changing economic geography of the traditional Chinese iron industry

Chapters 4 to 7 will describe what amount to four different iron industries. These used different technologies and operated in very different geographic and economic conditions. The iron industries of Sichuan, Shanxi, and Guangdong declined sharply in this period, while that of Dabieshan held its own and even prospered for a while. These developments become visible earliest in Guangdong, which was the first region to be affected by trade with the West, and we begin our investigation by looking at this trade.

Li Longqian lists four reasons for the decline of the Guangdong iron industry: intervention by the Qing state, domination by the guilds, a tendency to move capital out of industry into land, and the dumping of cheap commodities by foreign imperialists.[1] I shall not attempt to deal with the first two of these factors: a good deal of very detailed investigation would be necessary before I could speak with any confidence on the ways in which they affected the industry, and whether the net result was positive or negative. The third will be taken up briefly further below, but it is the fourth factor which I believe was the most important.

The terminology used by Li Longqian, 'dumping' by 'imperialists', will seem to many readers tendentious, but let us remember that Britain was explicitly imperialist in this period, and that the implicit threat of armed intervention gave the English East India Company a distinct advantage in its trade with China.[2] 'Dumping' suggests the deliberate sale of commodities

---

1. Li Longqian (*1981*, pp. 366–368). A more positive view of foreign trade is given by Ding Richu & Shen Zuwei (1992).

2. On the European China trade in the 18th and 19th centuries, see especially Greenberg (1951); Fairbank (1953); Dermigny (1964a); Dermigny (1964b); Morse (1926–29); Chaudhuri (1985); Murphey (1972); Moulder (1977, pp. 98–127); Osterhammel (1989).

below cost with the intention of ruining competitors; this specific intention would seem to be difficult to prove, but we shall see that commodities often were sold below cost by European traders in China. Nevertheless the principal factor was simply that by the middle of the nineteenth century European ironworks were producing iron at a fraction of the cost of producing it in China. It is likely that China, up to about 1700, had the world's largest and most efficient iron industry, but about that time the British iron industry began the extraordinary sequence of technical improvements which brought the price of iron dramatically down and was a leading factor in the Industrial Revolution.

Western competition

As early as 1750 a French ship landed some 30 tonnes of iron at Guangzhou. French, Dutch, and Swedish ships occasionally landed both iron and steel in the following decades, usually selling it at a loss. Some iron was landed by the English East India Company in 1801 and 1805, and in 1807 'a trial lot of iron bars' sold in Guangzhou at a better price than expected. From 1811 iron appears to have been one of the normal commodities imported by the EIC, and by 1834, the year of the abolition of the EIC's monopoly, foreign iron appears to have become very important on the Guangzhou market.[3]

The commercial agent C. F. Liljevalch (1796–1870), in a report to the Royal Swedish Chamber of Commerce in 1847, devotes ten pages to iron and steel in China, and gives some price details.[4] He states, 'after the most careful investigations', that the cost of producing Chinese bar iron and transporting it from the hinterland to the city of Canton (Guangzhou) cannot be less than $2^{1/4}$– $2^{3/4}$ Mexican dollars per picul for second quality and $3^{1/4}$– $3^{3/4}$ for first quality. The Mexican dollar (the most important medium of exchange in China's foreign trade at the time)[5] was worth 4s 4d (£0·22) sterling, and the picul was $133^{1/3}$ English pounds (61 kg).

Liljevalch's 'careful investigations', and the resulting very precise cost figures, must be taken with a grain of salt, for he can hardly have had the opportunity to acquire the necessary technical and economic information for such an estimate. What is clear, however, is that the actual price of Chinese bar iron on the Chinese market in Guangzhou, which he must have known though he does not state it, was higher than these figures, which amount to £11 and £15 per ton respectively for the two grades. They may be compared with his figures for prices of European iron in Guangzhou:

---

3. Dermigny (1964a, pp. 197, 262–283, 367; 1964b, pp. 702–703) (eighteenth century); Morse (1926–29, vol. 1, p. 292 [French import in 1750], vol. 2, p. 357, vol. 3, pp. 1, 138 [trial lot in 1807], vol. 3, pp. 157, 174, 189, 205, 226, 242); Ball (1972, p. 3); Anon. (1834, pp. 463, 471).
4. Liljevalch (1848, pp. 117–126).
5. On the use of the Mexican dollar in China, see especially Hao Yen-p'ing (1986).

commodity	price ($ per picul)
wrought-iron hoops from imported cotton bales	$2 - 2^{1/2}$
English bar iron	$3^{1/4} - 3^{1/2}$
English nail rod	$4^{1/2} - 4$
Swedish bar iron	5

It is clear that foreign iron was already competitive with Chinese iron in Guangzhou. Liljevalch also states that the cost of shipping 10 tonnes of iron from England to Guangzhou, including freight, customs duties, etc., would be about £30. The price of bar iron in England in the 1840s was about £7 per ton;[6] a quick calculation shows that the import of English bar iron to Guangzhou could yield, as early as the 1840s, a profit as high as 50 per cent.

It is difficult to put this profit figure into a meaningful context, for statistics on net profit in the European China trade are rare and in any case rather artificial. Imports of cotton cloth to Guangzhou by the English East India Company, for example, usually were sold at what appears in the accounts as a *loss* – meaning only that the profit on the corresponding exports was less than what appears in the accounts.[7] The greatest problem for Europeans trading in China was 'laying down the dollar' – the Mexican silver dollar.[8] It was necessary to pay for tea and other exports with silver, but carrying silver to China was the least profitable way of providing it. It was much more efficient to carry European products which could be sold for silver; but it was difficult to find imports for which there was sufficient demand in China, and many products were tried at one time or another. It was the opium trade which finally stopped and then reversed the flow of silver from Europe to China, but not all ships to China carried illegal cargoes. Every ship to China carried some sort of cargo, to help in laying down the dollar and also to serve as a ballast. The most common ballast cargo was pig lead, but as the number of ships to China increased the market for lead was easily glutted.[9] Bar iron was a natural substitute, especially as further technical developments brought down even more the cost of iron production in the West.

The cost of iron production in Europe continued to fall, and at the same time, because of both technological and institutional developments, ocean

---

6. The price fluctuated wildly in this period, with a minimum of £4·75 in 1844 and a maximum of £9·75 in 1847. Here I have taken the average of the figures for the years 1840–49 given in *Abstract of British historical statistics*, Mitchell (1971, pp. 492–493).
7. Dermigny (1964b, pp. 720–722).
8. Morse (1922); Cheong (1965).
9. Dermigny (1964a, p. 199); Morse (1922, pp. 233, 239).

freight rates also fell.[10] According to the Chinese Maritime Customs returns, China imported over 7,000 tonnes of iron in 1867, the first year for which statistics are available.[11] Two years later, in 1869, about 27,000 tonnes were imported. In 1891 the figure was 112,000 tonnes. Some of this imported iron supplied increased demand as China took its first steps toward industrialisation, but a large part, especially in the early years, simply replaced production in the traditional sector. About half of the imported iron was scrap,[12] for example old horseshoes. Scrap wrought iron was probably a fine material for Chinese smiths, and it was extremely cheap in the West.

The Qing government did not normally attempt to regulate commerce and industry directly; this would have required officials to have a detailed knowledge of some very technical activities. Instead it sold various kinds of monopoly or oligopoly rights to private individuals who then were held responsible for the ordered functioning of markets, the enforcement of the law, and the payment of taxes.[13] The most famous example of this means of regulation is the 'Thirteen Hongs' (*shisan hang* 十三行), the oligopoly which handled all foreign trade in Guangzhou.[14] Other examples are the salt gabelle and the Manchurian ginseng monopoly, and still another monopoly is the iron industry of Guangdong. It is not clear how this monopoly licensing system for the iron industry actually worked in detail, but it is fairly clear that state-granted monopolies of some sort did exist in the Guangdong iron industry.

It was a firm ideology, in fact an *idée fixe*, of the British traders in Guangzhou that all state regulation and all monopolies are pernicious. In 1842 the Treaty of Nanjing, the first of the unequal treaties forced upon China after its defeat in the first Opium War, contained a provision specifically banning the monopoly system.[15] Private trade monopolies on iron as well as other commodities were immediately formed, but the Qing state was forced to suppress these after complaints from British traders.[16]

---

10. Harley (1988).

11. Tegengren (1923–24, p. 400).

12. In 1899, the first year for which I have seen a breakdown of the import figures, scrap iron constituted 44 per cent of China's iron imports. Tegengren (1923–24, pp. 401–402); cf. Hosie (1901, p. 257). Hsiao Liang-lin (1974)'s normally very useful digest of foreign trade statistics does not, unfortunately, include information on imports of scrap and ordinary grades of iron.

13. This use of monopolies is discussed briefly by Dermigny (1964b, p. 64).

14. See e.g. Cordier (1902).

15. Spence (1990, pp. 158–160).

16. Fairbank (1953, pp. 306–307); Wakeman (1966, p. 97); Public Record Office, London, FO 228/51.

With the Treaty of Nanjing the Qing state was denied its only means of regulating the iron industry, and at the same time four more ports were opened to foreign trade. From this point on the decline of the iron industry became very rapid in Guangdong and began in other parts of the country.

Chapters 4–7 below will document the decline in the four regions in detail. There are also numerous other, more anecdotal, indications of decline caused by foreign trade, for example von Richthofen's description of Shanxi in 1870:

> The mining of coal, the manufacturing of iron, and the conveying of both to market employ a large number of men and animals. But notwithstanding its ample resources the country is poor. The profits are reduced to a minimum. ... Underground miners, who receive elsewhere 200 to 300 cash a day, must here content themselves with wages of 100 cash. Yet the owners of mines are poor people. There have evidently been better times in this region, as one is justified in concluding from the great number of houses built with luxury, and richly adorned with fine work of sculpture. It is possible that the introduction of foreign wrought iron, into those districts which are accessible by water from the Treaty ports, has greatly reduced the amount of sale and total production of Shansi iron, and that the desire to supply as many as possible of the former markets has tended to reduce the original price of the iron, and consequently the profits of the manufacturer.[17]

In the late nineteenth century Geerts mentioned the situation in China in his observations concerning Japan:

> Finally it may be noted that the manufacture of wrought iron in Japan has diminished considerably with the import to Japan of great quantities of iron in bars and plates, principally from England and Belgium. The convenient and diverse shapes of European wrought iron and their relatively moderate prices, together with the miserable state of the roads in the mining districts, are the causes which have made this metal an important article of foreign trade, in China as well as Japan, in spite of the abundance of excellent ores in both countries.[18]

Return on investment

While competition with cheap imported iron undoubtedly was the most important cause of the decline of the Chinese iron industry, other factors must also have been at work, for the decline of the Guangdong iron industry started in the eighteenth century, before significant amounts of iron began to be imported. An additional factor was probably that foreign trade brought

---

17. von Richthofen (1872, p. 31).
18. Geerts (1878–83, p. 540).

new investment opportunities for Chinese entrepreneurs, and that these investments could give a higher return than ironworks could.[19] Luo Yixing has noted a number of cases in which ironworks were closed down because ore deposits were worked out.[20] These are likely to be signs of a shortage of investment capital, for searching for a new deposit and thereafter opening up a new mine was expensive, and would have been undertaken only if the expected return was competitive with other possible investments. He believes the principal cause of the decline of the Guangdong iron industry to be that the province actually had no more rich ore deposits left; this seems on its face to be unlikely, and an economic explanation of the decline appears to be more credible.

China's first modern ironworks was established in 1891, in Hanyang, Hubei. In 1922 there were seven modern ironworks in operation. The vicissitudes of these enterprises are not part of the present story, however, for most of their production was sold to Japanese creditors at sub-market prices, while China continued to rely on the traditional sector and foreign imports for its own iron consumption.[21]

The effects of decline on technology

Competition with modern industry caused all of these regional industries to shrink, leaving fewer units and smaller total production; but the influence of this competition was not uniform over all ironworks. In fact it hit hardest precisely in the places where the most technically sophisticated and capital-intensive techniques were in use. The reasons are several. A prerequisite for a large highly-capitalised works with a large production is a large market, and this implies good transportation facilities;[22] but the regions with good transportation facilities were also the first to be penetrated by foreign goods. Furthermore, in China, capital was much more mobile than labour. As the profits of the highly-capitalised works declined because of falling prices the investors could move their capital into other, more profitable, enterprises, for example tea and opium. On the other hand the labourers, facing a continuously falling standard of living, seldom had much choice but to

---

19. For example Adshead (1984, pp. 110–111) notes a number of new investment oppor-
    tunities brought to Sichuan in the early twentieth century.
20. Luo Yixing (*1985*, pp. 90–92).
21. The story has been told by Tegengren (1923–24, pp. 365–397).
22. Obviously what I mean by 'good transportation' must be taken in relation to the par-
    ticular product involved. Transportation in Shanxi was by most measures dreadful,
    but the famous needles of Shanxi could have a very large market because transporta-
    tion was not a large part of their price.

continue producing iron. Furthermore, by 1900 at the latest, Chinese ironworks could no longer compete with foreign iron in quality, only in price.

The works that survived best were those in poor isolated regions like Dabieshan which produced for a purely local market and used labour-intensive low-capital methods. Their survival led to a curious phenomenon when World War I brought greatly increased prices for iron: the increased prices made the traditional methods viable again, but the best traditional methods had by this time been forgotten. The tiny blast furnaces of Dabieshan, which were appropriate for a small production for local markets, began to be used in mass production to supply a large part of southern Henan.[23] The inter-war depression in the West may also have had a positive effect on China's economy, and therefore on the traditional iron industry.[24]

The Great Leap Forward

These considerations have considerable relevance for the study of the campaign for iron production in the Great Leap Forward of 1958–59. The usual evaluation of that campaign, both in China and abroad, is that it was a total fiasco with no redeeming features.[25] Most contemporary accounts, even the wildly enthusiastic propaganda, tend to confirm this evaluation when they are read critically: there are very few signs that the thousands of 'backyard furnaces' actually produced any iron at all. Of the numerous photographs of traditional blast furnaces which can be seen in Chinese publications of the period, there are very few that show them actually in production. But according to a speech by Premier Zhou Enlai 周恩來 on 23 August 1959, in 1958 these primitive blast furnaces actually produced 4·16 million tonnes of usable pig iron (together with 4–5 million tonnes of pig iron of unusable quality).[26] That is, 30 per cent of the year's pig iron production (13·69 million tonnes of usable pig iron) was produced in these primitive blast furnaces which, in the opinion of most observers, were totally worthless. Many of the production statistics published in those years have later proved to have been greatly exaggerated: is this another example of the same?

It is more probable that the campaign actually was, to a certain extent, a success in those parts of the country where the traditional iron-production techniques had not been forgotten. Where production already existed for local purposes it could be expanded. This was normally the case only in places where transportation was bad. Here iron was produced using in-

---

23. Note the remarks of Rawski (1989, p. 249) on the effect of World War I on Chinese industrialisation.
24. Myers (1989).
25. E.g. MacFarquhar (1983); Liu & Yeh (1965, p. 115).
26. Anon. (1959b, p. 18). Note also the careful analysis of the industrial output statistics of the period by Subramanian Swamy (1973, pp. 41ff).

efficient methods and was therefore expensive, and the added cost of transportation made it even more expensive in the places where it was to be used; but it is quite possible that iron production was nevertheless an economically rational use of labour in isolated poverty-stricken regions.[27] The great error of the campaign was the attempt to re-introduce the traditional techniques in places where they were long forgotten, and where there also were better uses for labour.

It is rare that journalists, politicians, diplomats, or tourists travel in the poorest regions of China. Nearly all those who reported on the Great Leap Forward, both Chinese and foreigners, kept to places where travel was reasonably comfortable. The only exception I am aware of is Rewi Alley, who retained his contact with China's poor and travelled where few others had any desire to go.[28] He also had a fine feeling for what makes a good picture, and many of his photographs show blast furnaces in production. He notes proudly several times that it was the poorest peasants who produced the best iron: no doubt he felt that there were moral reasons for this, but we may note that there could very well have been economic reasons as well.

27. Herman (1956) and Ishikawa (1972) discuss the economic factors involved. Herman also shows that numerous Asian governments at this time were beginning to recognise the potential value of traditional small-scale industries.

28. Alley (1961a; 1961b). Charles Curwen writes: 'His experience of China, and his own character, led him – not surprisingly – to mistrust the rich, and he had a low opinion of the educated children of the rich and perhaps of intellectuals in general. He was prejudiced in favour of the poor and had a stubborn confidence in their natural ability and determination. This sentiment was confirmed by the quality of the young people, nearly all from poor often wartime refugee families, who were formed by the school, founded by the Chinese Industrial Cooperatives, of which he was director (and where I worked for about seven years). Later, in many different walks of life, they gained a reputation for their ability and their readiness to get their hands dirty.' (Letter to DBW, 30 March 1995).

3

Traditional Chinese iron production techniques

It will be useful to distinguish here between *primary* iron-production techniques, for the production of cast or wrought iron from ore, and *fabrication* techniques, those used by the ironfounders and smiths to make useful products from this raw material. It happens that the primary techniques differed greatly from place to place in China, while the fabrication techniques, to the extent that we can see them clearly in the sources, seem to have varied much less.

In the following we shall consider the traditional primary iron-production techniques of four parts of China: the Dabieshan 大别山 region of southern Henan and northern Hubei, and the provinces of Sichuan, Guangdong, and Shanxi. These places were chosen partly because of the availability of good documentation and partly because of the special interest of their technologies. The fabrication techniques will not be dealt with here.

The flow diagram of Figure 1 will serve to show the general structure of the traditional iron industry in the first three places mentioned. That of

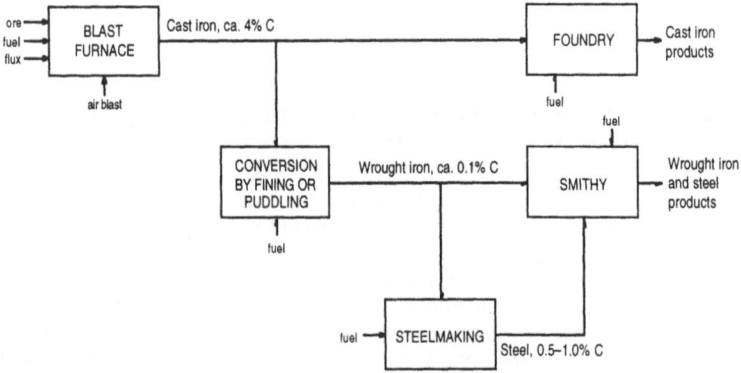

Figure 1. Flow diagram of traditional Chinese blast-furnace iron production

12

Shanxi, with its 'crucible smelting' technology, was quite different and will be discussed separately.[1] As in the modern steel industry, the process generally used in China was 'indirect'. Cast iron with a high carbon content was produced from ore in the *blast furnace* (see Box 1);[2] this product could be used directly in a foundry, but most of it was converted to wrought iron (or more correctly, mild steel)[3] with typically 0·1 per cent carbon. This was the basic material of the smith; when something harder was needed, for example for the cutting edge of a knife, it was necessary to put some carbon back into the iron, to make a medium-carbon steel. High-carbon steels, with over 1 per cent carbon, were rarely used. Some steelmaking techniques could be used by the smith himself, but he could also obtain steel stock from specialised producers.

The indirect process is the most efficient way of producing wrought iron, in spite of the curious roundabout way in which it works, with carbon first being put into the iron to make cast iron, then removed to make wrought iron, then put in again to make steel. The modern blast furnace is in principle not much different from the traditional Chinese blast furnaces, though it is much larger and has been improved in a variety of ways. The means by which carbon is removed are quite different, and in the West have changed a number of times, from the fining hearth of Medieval times (with some resemblance to the traditional Chinese fining hearth) to the puddling furnace, patented in 1784, to the Bessemer converter of 1855, and on to a variety of ever more efficient devices.

An important fact about indirect iron production is that it provides unusually large economies of scale: the greater the production, the lower the cost of the product per unit.[4] This fact has had enormous historical importance, and for example it is surely one of the factors in the rise of capitalism in the West. In China it may have been an important factor in the

---

1. Chapter 6 below.
2. The modern Chinese word is *gaolu* 高爐, which must come from the German *Hochofen*, probably through Japanese (Liu Zhengtan, *1984*, p. 114). A number of Chinese terms for 'blast furnace' have been used traditionally in particular localities, but there seems to have been no single widely-understood term for this particular type of furnace.
3. In Western contexts 'wrought iron' usually has approximately zero carbon, and the correct word for the iron produced in the traditional Chinese fining hearth would be 'mild steel' (which is defined as having less than 0·25 per cent carbon). In the traditional Chinese iron industry of recent centuries wrought iron in this sense was seldom or never produced, and it seems sensible to use the term in an extended sense, including low-carbon mild steel which was used for the same purposes as wrought iron in the West.
4. 'There are inherent and intrinsic economic differences between large outputs as compared with smaller outputs... The advantages are wholly in favour of large outputs.' Sidney (1920, p. 129).

rise of the state of Qin in the third century BC.[5] It is therefore curious, and in need of explanation, that the Chinese blast furnaces which we shall see in the following are found in such a range of sizes, from the 'dwarf' furnaces of Dabieshan, only 2–3 metres high, to the very large furnaces of Sichuan and Guangdong, up to 10 metres high. The explanation lies in the fact that large-scale production requires heavy investment and a large and stable market, so that potential economies of scale can be exploited only in regions with good transportation. In more isolated regions, such as Dabieshan, transportation costs added so much to the cost of iron that it was economically rational to set up a small-scale production for local needs in spite of its relative inefficiency. And these small furnaces are not all that inefficient: though at first sight they look 'primitive', it is probably more correct to see them as a highly sophisticated development out of the larger furnaces, providing reasonable efficiency on a production scale which fills the needs of the population of an isolated region.

---

5.   As I have argued in Wagner (1993, pp. 408–409).

4

Small-scale ironworks of the mountains of Dabieshan

The Dabieshan 大别山 range, around the point at which the provinces of Anhui, Henan, and Hubei meet, is a region of rugged mountains and fertile valleys where poverty is severe.[1] Until the 1960s there was hardly a road here, and most transport within the region was on foot or horseback. It has also traditionally been very isolated from other regions, for there is no good water transport available. The Beijing–Hankou Railway, completed in 1906, touches the region only at its extreme western end, at Xinyang 信陽, and seems to have had little impact on the local economy. The principal natural resources here are forests and minerals, but the isolation of the region has made any large-scale exploitation of these unprofitable. A small-scale iron industry has, however, been important in the local economy, and this survived at least until the Great Leap Forward of 1958–60, when it was the model on which many other regions based their attempts to build up small-scale ironworks.

In a small book of mine on the iron industry of the Dabieshan region,[2] I drew on two accounts by travellers who visited there, the Swedish geologist E. T. Nyström about 1916 and the Chinese geologist Guo Yujing 郭玉璟 in 1932, and on several technical studies prepared in connection with the Great Leap Forward.[3] My late friend Prof. Zenshirō Hara, in a review of that book, pointed out several other descriptions in local gazetteers,[4] and I have also found a few more descriptions.[5] With this material it is now possible to give an account of the technology in its geographical and economic context. We start with the technology.

1. Very little has been written on the human geography of this region, but see McColl (1967); Di Xianghua (1987); Li Runtian (*1987,* pp. 328–340).
2. Wagner (1985).
3. E. T. Nyström in Tegengren (1923–24, pp. 179–180, 334–335); Zhang Youxian and Guo Yujing (*1932,* pp. 239–241); Anon. (*1958a; 1958b*); Liu Zhichao and Tang Youyu (*1959*).
4. Hara Zenshirō (*1991*).
5. A recent Chinese study, not yet published, has been briefly reported by Miao Chang-xing and Li Jinghua (1994).

Figure 2. Blast furnace in Xinyang, Henan, photographed ca. 1916 by E. T. Nyström. Reproduced from Tegengren (1923–24, vol. 2, plate 34 left).

The blast furnace

The ore used in all these ironworks was ironsand, washed from river sand in sluices. This was a very rich ore, with iron content from 49 to 65 per cent according to different reports.[6] The richest ironsand contained about 90 per cent iron oxides and only about 5·5 per cent silica; with so little silica in the charge, blast furnace operation was greatly simplified, since a flux could be dispensed with.[7]

6. Respectively Nyström in Tegengren (1923–24, p. 180); Anon. (*1958a*, p. 6), tr. Wagner (1985, pp. 12, 49).

7. The exotic character of the iron-production technology of this region caused reviewers considerable confusion. Bennet Bronson (1987, p. 97) claims that ore with 65 per cent iron is impossible, and William Rostoker (1987, p. 347) claims that blast furnace operation without a flux is impossible. Neither reviewer explains the reasons for his disbelief, but both are mistaken.

Figure 3. Blast furnace in Xinyang, Henan, photographed ca. 1916 by E. T. Nyström. Reproduced from Tegengren (1923–24, vol. 2, plate 34 right).

The small blast furnace in which this ironsand was smelted was of roughly the same type throughout the region. Photographs of three are shown in Figures 2–4; the first two were taken in Henan in 1916, the third in Anhui in 1958. They are only about 2·2 metres high, and are built almost entirely of locally-available materials. The instructions for building the Huang Jiguang Furnace[8] 黃繼光爐 used in Macheng 麻城 County, Hubei, diagrammed in Figures 5–6, indicate that the walls of the furnace, which are 10–13 cm thick, are made of a mixture of loess soil, sand, and straw, reinforced with iron bands. This is then lined with a more refractory material, which contains 60 per cent finely powdered charcoal. Around the taphole, where the highest temperatures are encountered, blocks of sandstone are used.[9] In other reports it appears that these 'taphole stones' were the only part of the furnace which could not be obtained locally, but were brought in from as far as 200 km

---

8.   This furnace was named for Huang Jiguang (1930–52), a hero of the Korean War.
9.   Anon. (*1958c*).

Figure 4. Blast furnace in Jinzhai 金寨, Anhui, ca. 1958. Reproduced from Anon. (1959).

away.[10] The taphole was kept constantly open, since plugging it would have required refractory clay, which was not available locally; slag and iron were tapped by tilting the entire furnace, as can be seen in Figure 4. The cross-beams seen in Figures 2 and 4, and the chain in Figure 3, were used to limit this tilt.

The fuel used was charcoal, and blast was provided by a traditional 'windbox' (*fengxiang* 風箱, double-acting piston bellows). The furnace could operate continuously for 6–7 days before it was necessary to repair the inner lining and replace the taphole stones. In this time charcoal and ironsand were charged and molten iron tapped several times per hour. Reports from different times and different places in the region indicate that daily production of pig iron was between 0·6 and 1·2 tonnes, and that the amount of charcoal needed to produce one tonne of pig iron was between 2 and 7·5 tonnes.[11]

---

10. Guo Yujing reported in 1932 that in Xinyang the taphole stones were of sandstone from Jiayu 嘉魚, Hubei, while in Shangcheng 商城 they were of diatomite ('diatomaceous earth') from Qishui 蘄水, Hubei. Zhang Youxian & Guo Yujing (*1932*, p. 240); cf. Wagner (1985, p. 50).

11. The latter figure is that given by Nyström, but Tegengren (1923–24, p. 334 fn.), quoting it, considers it 'fantastically high'. It is quite possible that Nyström was misinformed.

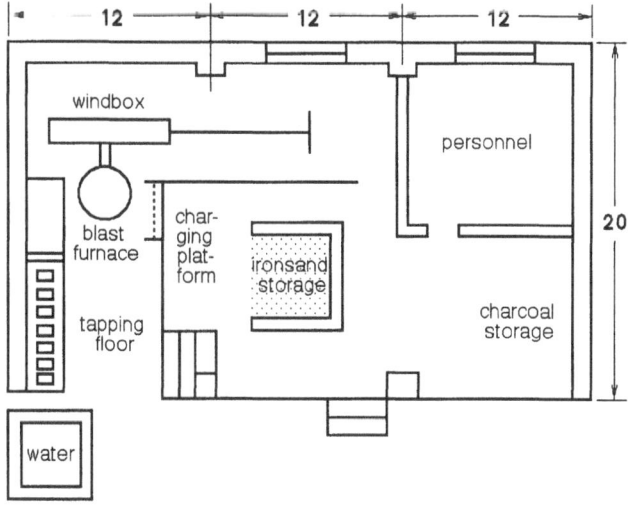

Figure 5. Plan of a 'smelting house' at East Wind People's Commune in Muzidian, Macheng County, Hubei 麻城木子店東風人民公社, redrawn from Anon. (*1958c*, p. 22). Dimensions are given in 'market feet' (*shichi* 市尺, 33 cm).

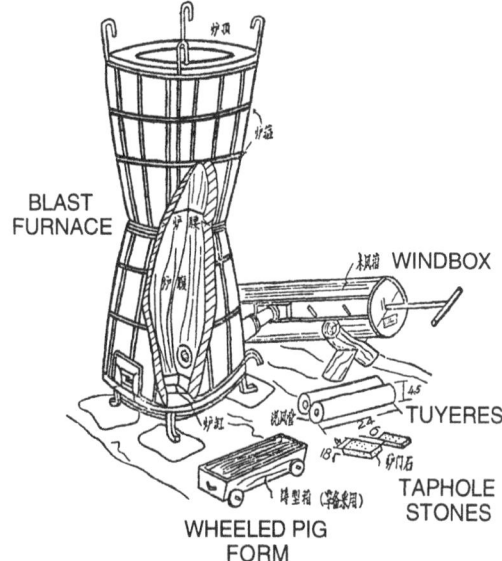

Figure 6. Sketch of the Huang Jiguang blast furnace in Macheng, Hubei, as used in the Great Leap Forward period, and associated requisites, reproduced from Anon. (*1958c*, p. 22). Dimensions are given in 'market inches' (*shicun* 市寸, 3·3 cm).

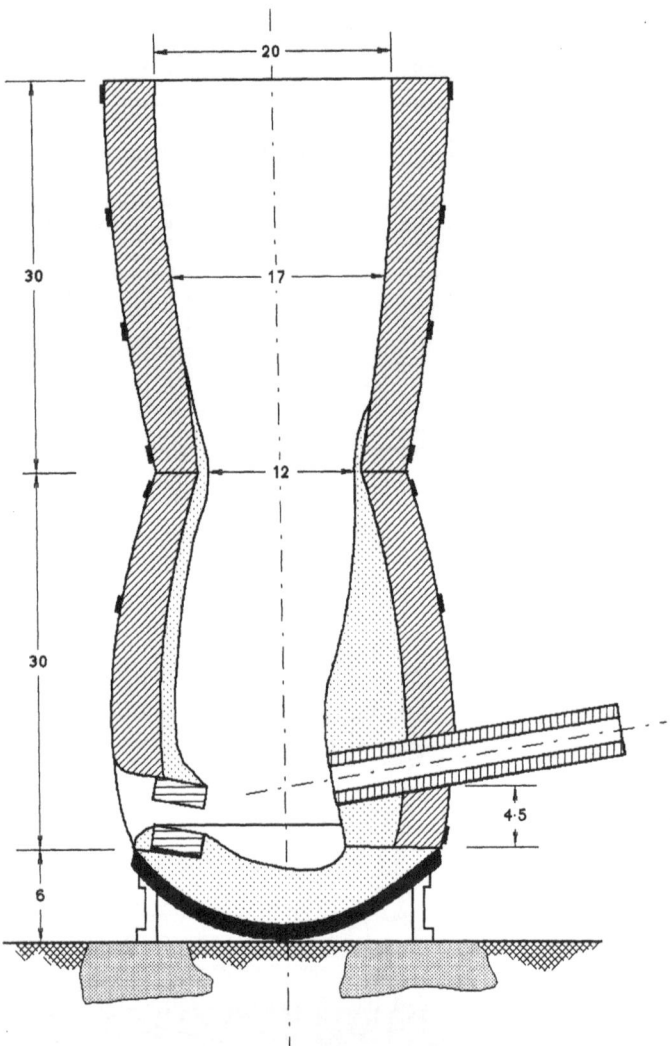

Figure 7. Vertical section of the Huang Jiguang blast furnace in Macheng, Hubei, as used in the Great Leap Forward period, redrawn from Anon. (*1958c*, p. 22). Dimensions are given in 'market inches' (*shicun* 市寸, 3·3 cm). Cf. Figures 5 and 6.

In a blast furnace the combustion of the fuel maintains a high temperature, at least 1200° C near the bottom, and also provides an atmosphere with a large concentration of carbon monoxide (CO). In this highly reducing atmosphere iron oxides are reduced to metallic iron. This iron takes up carbon, reaching a maximum carbon content in the range 4–5 per cent; at this carbon content the melting point of the iron is as low as 1147° C, and it melts. The molten iron collects at the bottom of the furnace until it is tapped. Box 1 gives a more detailed explanation of the operation of a blast furnace.

The ironsand used here contained about 5·5 per cent silica (SiO_2). It was necessary to get the silica out of the furnace in a free-flowing molten slag: silica itself has a very high melting point (over 1700° C), but a mixture of silica and wustite (FeO) can have a melting point as low as 1177° C.[12] The furnace is therefore arranged to have an oxidising zone near the bottom, where a small amount of iron is re-oxidised to FeO which mixes with the silica to form a reasonably low-melting slag.[13] Probably about 9 per cent of the iron in the ironsand was thus oxidised and lost in the slag; this was the cost of not using a flux, which presumably was not available locally. Limestone ($CaCO_3$) can form a slag with silica, and is one of the commonest fluxes used in modern blast-furnace operation.[14] Nyström in 1916 described an ironworks in Xinyang, Henan, which used a much less well-concentrated ironsand, with 49 per cent iron and 26 per cent silica; with so much silica a flux would have been advisable to avoid too much loss of iron, and he notes that limestone is available in the neighbourhood of the ironworks.[15]

The fining hearth

Cast iron from the blast furnace contains about 4 per cent carbon. It can be used directly by a foundry, but if it is to be used by a smith most of the carbon must be removed by a process which in Chinese is called *chao* 炒. This is an apt word for the process, for its usual meaning is 'stir frying', and the removal of carbon from cast iron involves carefully stirring about lumps of very hot iron. Various English words have been used to translate *chao*, including 'roasting', 'refining', 'converting', and 'puddling', but for the process used in the Dabieshan region I have chosen to use the nineteenth-century word 'fining', defined as 'the operation of converting cast into malleable iron ... in a hearth or open fire, urged by a blast of air with charcoal as the fuel.'[16] Further below, discussing the more sophisticated process used in Sichuan, I shall translate *chao* as 'puddling'.

---

12. Muan & Osborn (1965, p. 62, fig. 45a); Rosenqvist (1974, p. 345).
13. Miao Changxing & Li Jinghua (1994) report one slag analysis as follows:

constituent	SiO_2	CaO	TiO_2	Al_2O_3	FeO	K_2O	MgO	P_2O_5
per cent	38·47	17·95	18·15	10·74	7·09	3·24	2·43	1·57

(This gives a total of 99·64%.)
14. E.g. Peacey & Davenport (1979, p. 7).
15. Tegengren (1923–24, pp. 180, 335).
16. Percy (1864, p. 579). See also the long footnote on this subject in Wagner (1993, pp. 290–291, fn. 37).

Box 1. Technical details of the operation of an iron blast furnace.

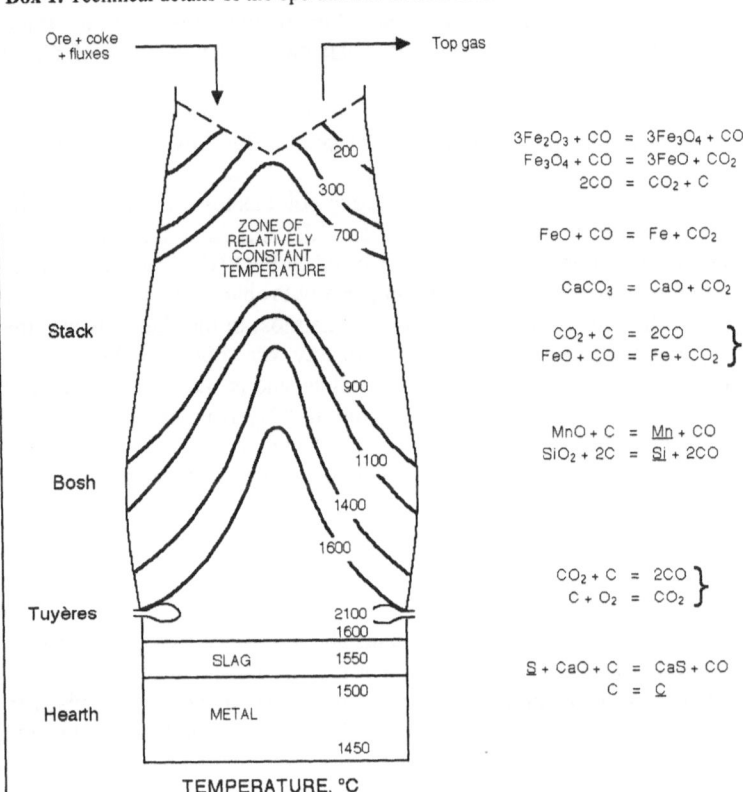

$$3Fe_2O_3 + CO = 3Fe_3O_4 + CO_2$$
$$Fe_3O_4 + CO = 3FeO + CO_2$$
$$2CO = CO_2 + C$$

$$FeO + CO = Fe + CO_2$$

$$CaCO_3 = CaO + CO_2$$

$$\left. \begin{array}{l} CO_2 + C = 2CO \\ FeO + CO = Fe + CO_2 \end{array} \right\}$$

$$MnO + C = \underline{Mn} + CO$$
$$SiO_2 + 2C = \underline{Si} + 2CO$$

$$\left. \begin{array}{l} CO_2 + C = 2CO \\ C + O_2 = CO_2 \end{array} \right\}$$

$$\underline{S} + CaO + C = CaS + CO$$
$$C = \underline{C}$$

TEMPERATURE, °C

Figure 8. Diagram of a modern large-scale coke-fuelled iron blast furnace, after Peacey & Davenport (1979, p. 17, fig. 2.1) and Rosenqvist (1974, p. 274, fig. 9-4). The height is typically 20–30 metres. Temperatures in °C are indicated for isotherms inside the furnace. The reactions which take place are indicated at the right; underlined elements are in solution in iron.

Empirical research provides considerable detail on what happens inside the modern blast furnace, and that is what will be described here. Of large charcoal blast furnaces like those traditionally used in Sichuan it is possible to say that the basic principles are the same but with the major difference that all temperatures are lower. Of small charcoal blast furnaces like those of Dabieshan it is reasonable to assume that the basic principles are approximately the same, but with the possibility of fundamental differences, of which we know, at the moment, nothing.

A modern blast furnace operates continuously for months or years at a time, with **coke**, **ore**, and **flux** being charged in the top, **air** being blown through numerous tuyères near the bottom, and molten **iron** and **slag** being tapped out of tapholes at the bottom. Operation continues until the furnace lining has been so damaged by the high temperatures that it is necessary to repair it.

The ore has been calcined (roasted) before charging, so that the iron in it is entirely in the form of Fe_2O_3 (ferric oxide, hematite). The fundamental reactions are the reduction of this by CO (carbon monoxide), first to Fe_3O_4 (ferrosoferric oxide, magnetite), then to FeO (ferrous oxide, wustite), and finally to metallic iron. Some carbon, typically ca. 4 per cent by weight, dissolves in the iron near the bottom; with this carbon content the melting point of the iron is under 1200°C.

The combustion of coal at the tuyères produces CO_2 (carbon dioxide), and this reacts with carbon to produce the necessary CO. The CO reacts with the iron oxides to produce CO_2 again, this reacts with carbon to produce CO, and so forth in a cycle. The necessary conditions for each reaction are diagrammed in Wagner (1993, p. 388, fig. 7.53) or in any textbook of engineering thermodynamics. What is important is that high temperatures and very high concentrations of CO in the furnace atmosphere are required for the reduction of FeO. The necessary concentration of CO is more readily obtained with charcoal as the fuel than with coke (because charcoal is more reactive), and therefore charcoal blast furnaces operate at lower temperatures.

Iron smelting would be easy if the ore were composed of nothing but iron oxide. In fact all ores contain significant amounts of SiO_2 (silica) as well as other minerals: this unwanted material is collectively called the **gangue** of the ore. If the furnace is to operate continuously the gangue must be removed in molten form, but it will normally have a melting point which is much higher than the temperatures required for the reduction of iron oxides. Therefore a **flux**, typically $CaCO_3$ (limestone), is charged along with the ore and the fuel. The flux is chosen to form with the gangue a **slag** with a practically low melting point (less than 1400°C in modern practice). $CaCO_3$ decomposes in the furnace to CaO (lime) and CO_2, and the mixture of CaO and SiO_2 has a much lower melting point than either mineral alone. Other minerals in the charge, either incidentally present in the gangue or intentionally added in the flux, may further depress the melting point of the slag. Especially important here is Al_2O_3 (alumina).

Limestone is not only an excellent flux, it also has the property that it can remove sulphur (S) from the liquid iron by the reaction shown. In modern practice, using coke with fairly high sulphur, large amounts of limestone are used, giving a CaO/SiO_2 ratio in the slag as high as 1·2; in charcoal-fueled blast furnaces sulphur is rarely a problem, and much smaller amounts of limestone are used.

In some pre-modern Chinese blast furnaces no flux is used. It seems that in these cases one or more of several special conditions must hold: (1) the ore used may be very rich, i.e. contain only a small amount of gangue; (2) the ore may be 'self-fluxing', the gangue containing significant proportions of limestone or alumina; (3) the charcoal may be from a wood which has grown on chalky ground and therefore contains a significant proportion of lime; (4) in operation enough of the furnace lining may be consumed to add significant amounts of alumina or other minerals to the slag; (5) the internal form of the furnace may be arranged in such a way that a small amount of the reduced iron is re-oxidised to FeO near the bottom, providing a very effective flux for silica. In pre-modern furnace operation it also appears possible to tolerate a slag which is rather viscous and does not separate well from the iron: the product is a 'slaggy' iron which iron-founders, finers, and puddlers nevertheless are able to use without great problems.

The hearth in which fining was done in Shangcheng 商城, Henan, in 1958 is diagrammed in Figure 9, and a pair of similar fining hearths in Xinyang in 1916 is sketched in Figure 10. (Fining hearths were normally built in pairs in this region because alternating between the two helped to save the refractory lining and give each a longer effective life.) Wood, charcoal, and broken pieces of cast iron were charged into the hearth and ignited; air was pumped in, and the charge was stirred about with an iron rod. When the carbon content of the iron was sufficiently reduced it was removed in small balls which were hammered to remove slag and form them into bars. I have described the process in more detail elsewhere.[17] I have not seen a photograph of a fining hearth in use in the Dabieshan region, but Figure 20, which shows a somewhat different fining hearth in use in Shanxi in 1958, will give a general impression of the operation. According to a description of the fining operation as practised in Shangcheng about 70 kg of wrought iron were produced in one fining cycle, and there were eight or nine fining cycles in a 12-hour shift. To produce one tonne of wrought iron the inputs were about 1·2 tonnes of cast iron, 85 kg of wood, and 100 kg of charcoal; the labour used was about 170 worker-hours.[18]

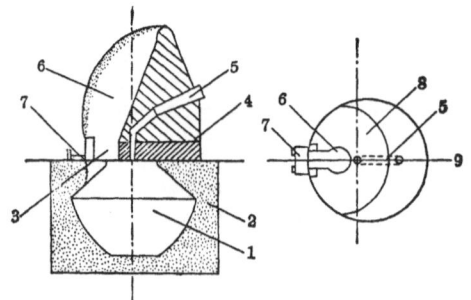

插图 60　河南省商城县土法炒钢炉的正视图和剖视图
（采自冶金工业出版社编《土铁土法炼钢》1958 年 10 月版，
第一篇《河南商城县的土法低温炼钢》）
部位名称：1.炉缸　2.夯实的耐火泥　3.炉门口　4.天门盖　5.通风管
6.炉门　7.炉门铁　8.炉窝　9.地面

Figure 9. Diagram of a fining hearth used in Shangcheng, Henan, in 1958, reproduced from Yang Kuan (*1982*, p. 225); orig. Anon. (*1958b*, p. 23). **1**. Hearth. **2**. Tamped fireclay. **3**. Hearth opening. **4**. Cover. **5**. Blast pipe. **6**. Hearth opening. **7**. Iron reinforcements. **8**. Nest. **9**. Ground level.

Temperatures of at least 1400° C must have been reached in this hearth. In the fining operation carbon in the iron was oxidised both by the oxidising

17. Wagner (1985, pp. 22–26, 60–67).

18. Anon. (*1958b*, pp. 11–16); Wagner (1985, pp. 60–66).

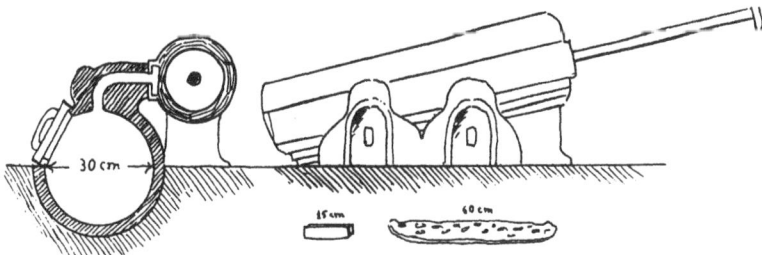

Figure 10. Drawing by E. T. Nyström of a pair of fining hearths in southern Henan, ca. 1916. In the background a traditional Chinese 'windbox', in the foreground a plate of pig iron and a wrought-iron bar. Reproduced by courtesy of Tom Nyström and the Museum of Far Eastern Antiquities, Stockholm.

atmosphere in the hearth and by an oxidising slag.[19] The pasty lumps of wrought iron removed from the hearth were heavily intermixed with this slag; hammering the iron on an anvil 'squeezed' the slag out like water from a sponge.

The product

Some analyses of cast iron and wrought iron produced by traditional techniques in the Dabieshan region are:

		C	Si	Mn	P	S
Cast iron	1958*	4·29	0·198	0·257	0·345	0·027
	1958†	3·65	0·13		0·59	0·042
		3·9	0·06		0·26	0·019
		3·02	0·01		0·42	0·044
Wrought iron	1916‡	0·34	0·37		0·16	0·045
	1958**	0·101	0·096	0·06	0·206	0·00324

*   Anon. (*1958a*); Wagner (1985, p. 20, 25, 87 n. 18).
†   Liu Zhichao & Tang Youyu (*1959*).
‡   E. T. Nyström in Tegengren (1923–24, p. 336).
** Anon. (*1958a*); Wagner (1985, p. 25).

19. The slag consisted of FeO produced by the oxidisation of some of the iron together with SiO_2, CaO, and Al_2O_3 from the hearth lining. It may be supposed that the most important reactions in the fining operation were: $2Fe + O_2 = 2FeO$; $C \text{ (fuel)} + O_2 = CO_2\uparrow$; $C \text{ (in Fe)} + CO_2 = 2CO\uparrow$; and $C \text{ (in Fe)} + 2FeO = 2Fe + CO_2\uparrow$.

The sulphur content of this iron is very low, even compared with the best modern steel. The silicon content of the cast iron is also exceedingly low in comparison with the iron used in most modern iron foundries: this is a feature of most pre-modern cast iron.

The iron industry of Dabieshan

The Dabieshan region has never held much interest for Chinese historians, nor has the iron industry. These two factors naturally frustrate the search for information on the history of the region's iron-production technology. The first sign is two incidental mentions of iron-production activity in a seventeenth-century geographical work, Gu Zuyu's *Du shi fangyu jiyao* 讀史方輿紀要: he explains the names of Tielu Shan 鐵爐山, 'Iron Furnace Mountain', in Huoshan County 霍山縣, and Dalu Shan 大爐山, 'Great Furnace Mountain', in Susong County 宿松縣, by noting that there are iron smelters near these places.[20] Presumably there were iron smelters elsewhere in the region which neither Gu Zuyu nor the authors of his sources had any reason to mention.[21]

In local gazetteers there are signs of an upswing in this iron industry in the nineteenth century. In Yingshan 應山, Hubei, according to a gazetteer published in 1990,

> In the Jiaqing 嘉慶 period [1796–1820] a man surnamed Ai 艾, from Huangpi 黃陂 [Hubei] established at Xiadian in Jiangxidian 漿溪店下店 an iron smelter where iron woks were cast using handicraft methods, with the local ironsand as the raw material and charcoal as the fuel. Its products were marketed in the county seat and in Guangshui 廣水 and Sui 隨 Counties [Hubei], as well as in Henan. From that time, over a hundred years ago, its production has never stopped. ... [22]

And in Xinyang, Henan,

---

20. *Du shi fangyu jiyao*, ch. 26, pp. 8a, 12a. A Ming-period local gazetteer, published in 1584, mentions illegal mining in Huoshan and also in neighbouring Huoqiu 霍丘 County, but it is not clear whether this was iron mining (*Lu'an zhou zhi*, ch. 4, pp. 16a–16b).

21. There seem also to be mentions of iron-production in Macheng 麻城 and Huang'an 黃安 Counties in the early Qing period. Xia Xiangrong et al. (*1980*, pp. 165–168), give a table of places in all of China which were found, in a search of a long list of early Qing sources, to have iron production. Here Macheng and Huang'an are listed, but the specific sources which mention them are not indicated. A local gazetteer for Macheng County states that the *Du shi fangyu jiyao* mentions an 'iron mountain' here, but I have been unable to verify this (Yu Jinfang (*1935b*, ch. 3, p. 37b); cf. *Du shi fangyu jiyao*, ch. 76, pp. 24a ff).

22. Anon. (*1990*, p. 261). The passage continues with some further data on this factory as it was in the 1930s and 1940s.

> At the beginning of the Daoguang 道光 period [1821–50] a Cantonese man living in Xishuanghe 西雙河 noticed that the sand was rich in iron. He was the first to teach people the method of washing and smelting it; he established factories for [smelting] iron and [casting] woks, and made an annual profit of more than a hundred thousand taels of silver (*jin* 金). ... [23]

It is unlikely that either of these entrepreneurs actually introduced a new technology. More probably, they saw the potential of an existing small-scale industry, given capital investment and broadened marketing. Around this time the market price of iron in Hankou, not very far away, was much higher than in Guangdong; this was the conclusion of an official investigation in 1841.[24] It may be that this price differential was relatively new toward the beginning of the nineteenth century, and was the reason for the new entrepreneurial activity.

From the beginning of the twentieth century there are numerous mentions of a flourishing small-scale iron industry in the region. A gazetteer for Huoshan 霍山, Anhui, published in 1905, in a section on mining which uses a curious mix of modern and traditional terminology, states:

> There is no information on copper or tin here. There are many places with iron ore, but as yet few know how to recognise the outcrops. The iron produced within the county is made by washing sand and smelting it. The method of smelting is as follows. First the sand is blown [*shan* 煽] in a blast furnace [*gaolu* 高爐] and transformed to liquid [*zhi* 汁] which is tipped out [of the blast furnace – cf. Figure 4] to form plates [*wa* 瓦] of cast iron [*sheng tie* 生鐵]. This is used in casting bells and gongs, woks and pots, agricultural implements, and the like. The cast iron can [also] be charged into a furnace in the earth [*dilu* 地爐, cf. Figures 9, 10] and fined [*chaolian* 炒煉] to make wrought iron. Steel is purchased from Wuhu 蕪湖 [Anhui] or from overseas; the local people are not able to make it ...[25] But the quality of the iron is excellent, and in

---

23. Chen Shantong (*1936*, ch. 3, p. 6).
24. William T. Rowe (1984, pp. 74–75, 361 n. 72) tells of this investigation by Yutai 裕泰, governor-general of Hu–Guang 湖廣. 'Yutai had been ordered by the imperial government to procure iron at Hankou to be cast into cannon for use against the British in Guangdong. When he approached some local iron brokers as mediators for the transaction, the governor-general was aghast at the price the metal commanded on the market, and initiated an investigation. Yet the investigation proved to Yutai's satisfaction that the price was neither fixed nor artificially inflated, but was kept up merely by the mechanisms of supply and demand, as successfully mediated by Hankou's iron brokers.' Rowe cites a memorial by Yutai dated Daoguang 道光 21/11/28 in the Qing Palace Archives.
25. The elided passage is a brief discussion of Western steelmaking techniques, written in smaller characters than the rest of the text.

the subprefecture [Lu'an Subprefecture 六安州] many people are pleased to buy and use it. Because the waterways are shallow and impassable, shipping it is laborious and costly; therefore very little goes outside the borders.[26]

This last is what we should expect on ordinary principles of economic geography: in a region without adequate water transport, iron will be produced only for local use. But before long the Dabieshan region was producing iron on a large scale and marketing it over a wide area. Nyström estimated in 1916 that in the part of the region which lies in Henan about 100 ironworks were in operation, producing ca. 14,000 tons of iron per year, and he stated that this production was carried by coolies all over southern Henan.[27] Further statements to the same effect, though without numerical estimates of production, are found in several local gazetteers published in the 1920s and 1930s.[28]

It may safely be assumed that these small-scale ironworks in normal times produced only for local needs, but they acquired a wider importance in the early twentieth century. By the end of the nineteenth century competition with cheap foreign iron had ruined the iron industries of most regions, and China was largely dependent on imports for its iron. In isolated regions, however, the price of transport made the foreign imports more expensive than products of the local ironworks, and these were able to continue production. When World War I caused the price of European iron to rise, and especially after the American embargo on iron exports of January 1918, the price of iron in China rose catastrophically.[29] Generally, the traditional iron industries of less isolated regions had already succumbed to foreign competition several generations before, and their techniques had been forgotten. On the other hand, at the new prices it became profitable for the ironworks of Dabieshan to expand their production and sell iron well outside the region, transporting it on the backs of coolies for lack of cheaper means.

---

26. Qin Dazhang & Ho Guoyou (*1905*, ch. 2, pp. 28a–28b).

27. Tegengren (1923–24, p. 334, 336). He estimated that there were 15 ironworks in Xinyang 信陽 County, 10 in Guangshan 光山 County, and 75 in Shangcheng 商城 County. It is unfortunate that we do not have Nyström's report itself, but only Tegengren's brief summary, for we should like to know how these estimates were arrived at. We appear to have no serious estimates for iron production in the rest of the region, the parts which lie in Anhui and Hubei.

28. **Yingshan** County 英山縣: Xu Jin et al. (*1920*, ch. 1, p. 26a, ch. 8, pp. 11b–12a); **Qianshan** County 潛山縣: Wu Lansheng et al. (*1920*, ch. 4, p. 21b); **Susong** County 宿松縣: Yu Qinglan et al. (*1921*, ch. 17, pp. 9a–9b, ch. 18, pp. 1a–4a); **Macheng** County 麻城縣: Yu Jinfang (*1935b*, ch. 3, p. 37a); **Xinyang** County 信陽縣: Chen Shantong (*1936*, ch. 7, p. 16a–16b, ch. 12, p. 5b, 6a); **Guangshan** County 光山縣: Yan Zhaoping (*1936*, ch. 1, p. 8).

29. C. T. Huang (1919); Hou Defeng & Cao Guoquan (*1946*, p. 816); Zhu Sihuang et al. (*1948*, p. 283); Hu Boyuan (*1946*, pp. 799–800); Reardon-Anderson (1991, p. 271); Lu Manping & Jia Xiuyan (*1992*, p. 14).

5
Large-scale ironworks in Sichuan

The Red Basin of Sichuan is a hilly region of intense agriculture surrounded on all sides by high mountain ranges. Communication within the region is facilitated by the famous Four Rivers from which the province derives its name, but communication with the rest of China is difficult. The Yangzi River joins Sichuan to eastern China through a series of gorges, and the famously arduous 'Road to Shu' (*Shu dao* 蜀道) joins it to Shaanxi to the north. The fertility of the soil and the mildness and dependability of the climate make this one of the breadbaskets of China; it has attracted immigrants throughout Chinese history, and its population density is extreme.[1]

In 1872 Ferdinand von Richthofen, after defining the limits of the roughly triangular Red Basin, summed up the human geography of the region as follows:

> Within this triangle there is life, industry, prosperity, wealth, intercommunication by water. Outside of it, as a rule, no river is navigable, with the exception of the Yangtze where it leaves the basin. To the south and west commence immediately territories occupied by *I-jên* [*Yiren* 夷人] or 'barbarians,' and in every direction we ascend from the elevated region of the Red Basin into the rugged mountainous countries which surround it. From the basin is derived that large and valuable produce

---

1. On the physical and human geography of Sichuan see Richard (1908, pp. 104–119); Dautremer (1911, pp. 173ff); Anon. (1944, pp. 85–92); Wiens (1949); or any of the many available geographies of China. On the region's economic history, Kapp (1973); Smith (1988); Bramall (1993); Zhang Xiaomei (*1939*); Meng Xianzhang (*1943*); Zhou Kaiqing (*1972*); Chen Shisong & Jia Daquan (*1986*); Du Shouhu & Zhang Xuejun (*1987*); Zhang Xuejun & Zhang Lihong (*1990*). On its geology and mineral resources, von Richthofen (1872, pp. 115–134; 1877–1912); Abendanon (1906); Tegengren (1923–24, pp. 281–283); Way (1916); DuClos (1898); Lei Baohua (*1943*); Zhou Lisan et al. (*1946*, maps 52ff). Of numerous travel descriptions the most important for our purposes appear to be: Széchenyi (1893); Cremer (1913); Robertson (1916); Hosie (1922); Richardson (1945); Needham & Needham (1948).

which has justly attracted attention of late years. Outside of it, on all sides, the country is thinly inhabited and little productive.[2]

These geographical considerations mean that there are good conditions for a local iron industry here: the demands of a large population, excellent intra-regional transportation, and isolation from the iron industries of other regions. The traditional salt industry of Sichuan consumed enormous numbers of very large salt-boiling pans, and this extra demand, over and above the normal iron consumption of a dense agricultural population, made from early times for a very large iron industry. In recent centuries the Sichuan iron industry used the largest blast furnaces to be found anywhere in China.

The iron industry has in recent centuries been concentrated in two parts of Sichuan: at the edge of the Red Basin southwest and south of Chengdu, and in the mountains along the Yangzi.[3] Some iron production is also reported in and near the mountains north of Chengdu. Iron ore – mostly clay ironstone – is very widely found in sufficient quality and quantity for production on a the scale of the pre-modern iron industry, and the location of iron production would seem to be determined more by the need for water transportation of raw materials from the mines and forests and of finished products to consumers. In particular the concentration of salt production south of Chengdu meant a concentrated demand for cast-iron salt-boiling pans in the same region.

The technology of this iron production is again as in Figure 1, but in comparison with the Dabieshan iron industry the scale of production was much larger, and water power was often used to power the blast of the blast furnaces. Detailed descriptions are available from 1877, 1936, the Second World War, and the Great Leap Forward;[4] there are also brief descriptions in local gazetteers and by many travellers.[5] In addition it seems that the technology of iron production in Sichuan has much in common with those in Yunnan and in Hunan, and there are a number of published descriptions of these.[6]

---

2. von Richthofen (1872, p. 115).
3. Xia Xiangrong et al. (*1980*, p. 167); von Richthofen (1872, pp. 123–124); Tegengren (1923–24, pp. 281–283); DuClos (1898, pp. 311–314); Zhou Lisan et al. (*1946*, map 55).
4. Széchenyi (1893, pp. 678–679); Luo Mian (*1936*, pp. 18–35); Zhang Xiaomei (*1939*, pp. Q13–Q14); Zhu Yulun (*1940*); Wang Ziyou (*1940*); Hu Boyuan (*1946*, pp. 800–801); Anon. (*1958d*; *1960*); Li Renkuan (*1959*); Zhang Chengji (*1959*).
5. Chen Buwu a.o. (*1928*, ch. 13, pp. 9a–11a); DuClos (1898, pp. 313–314); Cremer (1913, *passim*); Robertson (1916, p. 269); Way (1916, pp. 22–23).
6. **Yunnan**: Huang Zhanyue & Wang Daizhi (*1962*); Rocher (1879–80, vol. 2, pp. 195–218); Moore-Bennet (1915, pp. 220–221); Coggin Brown (1920a, pp. 82–97); Coggin Brown (1920b, pp. 337–339); Tegengren (1923–24 pp. 347–364). **Hunan**: Tegengren (1923–24, Chinese pp. 234–236, English pp. 338–339); von Richthofen (1877–1912, vol. 3, pp. 455–456); Lux (1912); Mao Zedong (1990, pp. 105–107); Yang Kuan (*1960*, pp. 135–137).

The blast furnace

The earliest description of blast furnace iron smelting in Sichuan[7] appears to be that of the Hungarian traveller Béla Széchenyi,[8] who visited an ironworks about 150 km southwest of Chengdu in 1877. His description is translated in Box 2.

This is one of several early descriptions of ironworks in Sichuan which indicate that the blast was water-powered.[9] The latest description of water-powered blast in Sichuan refers to observations in about 1915; after that only human-powered blast is mentioned.

Széchenyi notes that the ore used here is 'blackband', an ore consisting largely of siderite (ferrous carbonate, $FeCO_3$), which has a theoretical iron content of 48·2 per cent.[10] His estimate of 40–60 per cent iron in the ore is therefore over-optimistic,[11] but it does indicate that a very rich ore was used. The 'calcining' or roasting of the ore, seen at the extreme left of Figure 11, serves to drive off water, to convert hydroxides and carbonates to oxides,[12] to eliminate sulphur,[13] and to make the rock more porous and friable.[14] The calcined ore would be charged into the top of the blast furnace together with charcoal as the fuel and limestone as a flux. The flux serves two purposes: to form with the gangue of the ore a free-flowing slag, and to remove to the slag

---

7. We may note here the possible existence of an earlier description. The 1928 edition of the local gazetteer for Dazhu County 大竹縣 contains a curiously garbled description of blast furnace iron smelting which seems to be based on a much older description, edited by a person with modern technical knowledge (Chen Buwu *1928*, ch. 12, pp. 3a–3b; ch. 13, pp. 10b–11a). It would be very valuable to find this older description.

8. On Széchenyi and his expedition see Kreitner (1881) and an obituary in Hungarian by L. Lóczy (1923), which includes a long bibliography. Dr. László Ottovay of the National Széchenyi Library, and Dr. Csaba Horváth of the Hungarian National Museum, both in Budapest, have informed me that most of the material collected by the expedition was destroyed in Count Széchenyi's manor house in Nagycenk during World War II. The paleontological, mineral, and zoological materials, in museums in Budapest, were destroyed in 1956. What survives is the botanical material, in the Hungarian National Museum.

9. See e.g. DuClos (1898, p. 313); Cremer (1913, p. 58); Way (1916, p. 22); Tegengren (1923–24, p. 344).

10. Rostoker & Bronson (1990, p. 42).

11. DuClos (1898, p. 312) describes a deposit of sideritic ore near Chongqing with 35–40 per cent iron: still a very rich ore. Note also Tegengren (1923–24, p. 281)..

12. By such reactions as $FeCO_3 = FeO + CO_2 \uparrow$.

13. By such reactions as $2FeS_2 + 5·5 O_2 = Fe_2O_3 + 4SO_2 \uparrow$. Rosenqvist (1974, p. 245).

14. See e.g. Percy (1861, pp. 19–20); Rosenqvist (1974, pp. 238–259). In Széchenyi's description the calcination was done in an open heap, but descriptions of other ironworks indicate that it was done in a special stall furnace. See e.g. Luo Mian (*1936*, p. 18).

Box 2. Description of a blast furnace in Sichuan by Béla Széchenyi (1893, pp. 678ff). Cf. Kreitner (1881, pp. 805–806); Tegengren (1923–24, p. 342). Dr. Katalin T. Biro of the Hungarian National Museum has kindly corrected my translation from German against the Hungarian original, Széchenyi (1890, pp. 606ff)

Hoani-pu [i.e. Huangnipu] lies in a narrow valley, and bears throughout the stamp of a typical mining district. Everything here is black with coal dust from coal mining and iron industry.

In the neighbourhood chain bridges cross over the streams. Coal and iron occur together in the immediate vicinity. On the opposite bank is a blast furnace, ca. 8–9 m high and 5·5–6 m broad at the base. In form it is quite similar to a European blast furnace; it is built of stone and held together by an external wooden construction.

The blast is provided by a [piston-]bellows 1 m in diameter and 3·5 m long. This bellows, or rather, cylinder, is constructed on the same pattern as the common Chinese kitchen blowing cylinder, except that the piston is driven by a water-wheel.

For the tapping of both ore [i.e. iron] and slag there is only one opening, at least I did not see a special opening for the latter.

The ore smelted here is an ironstone (blackband), with 40–60 per cent iron, which occurs between the coal measures. Next to the shaft of the blast furnace the ore is first mixed with charcoal and roasted. As the works stood empty and out of operation, I was unable to obtain further data on the smelting process. I give the construction of the furnace in [Figure 11].

In the storehouse of the works was a large number of cast iron slabs measuring 1 m long, 0·60 m broad, and 0·02 m thick. The surface of this cast iron is very slaggy because of the lack of a separate outlet for the slag; its fracture is steel-grey and full of blow-holes throughout. Next to the blast furnace was the foundry, which however was also out of operation. ...

Figure 11. Sketch and sections of a water-powered blast furnace at Huangnipu 黃泥鋪 in Rongjing County 榮經縣 (modern Yingjing 榮經), Sichuan, ca. 1877, reproduced from Széchenyi (1893, p. 678, figs. 116–118). Height 8–9 m, base 5·5–6 m.

some of the remaining sulphur in the furnace charge (ore and fuel).[15] The Scottish mining engineer R. Logan Jack visited an ironworks in the same place, and his description specifically mentions the use of limestone:

> ... The works turned out to be not only a foundry, but also a smelter, operating on hæmatite, limonite, and a clay-band ironstone, the latter of which had been calcined at the mine. We were informed that the grade was 40 or 45 per cent – of course on the basis of extraction [rather than laboratory analysis]. A quantity of limestone was stored for flux. At the time of our visit the furnace was not in blast, all hands being busied on the conversion of the pig-iron into pots in the foundry. The blast ... was furnished by a turbine and wooden, double-acting cylinder of considerable size. The furnace was about 30 feet [9 m] in height, [with inside diameter] 5 feet [1·5 m] at the tuyères, and 10 feet [3 m] at the boshes [the widest part], and was fed through a very small opening at the top. It was built in part of hewn stone, and was not unlike the old English charcoal furnaces. The iron was cast into plates about $4 \times 2 \times 1^{1/4}$ inches [$10 \times 5 \times 3$ cm], and was fine-grained, and appeared to be of good quality. After breaking up the plates, the iron was melted in small cupolas with hand-bellows, and carried in iron hand-barrows to the casting department.[16]

Curiously, I have been unable to find any other early description of a blast furnace in Sichuan which explicitly mentions the use of limestone as a flux. Several descriptions, in fact, explicitly state that limestone was not used.[17] An example is the geologist L. Cremer's description of an ironworks in 1905 in the southern part of Nanchuan County 南川縣, which also gives other useful details:

---

15. Percy (1861, pp. 18ff); Percy (1864, pp. 349–350); Rosenqvist (1974, pp. 392–395); Peacey & Davenport (1979, pp. 7–8, 179). See Box 1.
16. Jack (1904, pp. 93–94).
17. A 1940 survey states that traditional ironworks in Sichuan 'do not add, or add only small amounts of, limestone.' It points out that this practice means that the slag produced is acid and does not attack the sandstone lining of the furnace as severely as a basic slag would (Wang Ziyou, *1940*, pp. 2, 3, 4). It gives the following slag analyses for samples of slags from ironworks in two counties:

County	CaO %	MgO %	SiO_2 %	Al_2O_3 %	S %	Mn %	Fe %
Weiyuan 威遠	7·85	5·80	45·63	18·22	0·12	8·04	8·74
Qijiang 綦江	5·66	1·35	42·66	9·37			6·32

The silica (SiO_2) would have come largely from the ore, and was the iron-smelter's greatest problem. The lime (CaO), magnesia (MgO), and alumina (Al_2O_3), since they did not come from a flux, could have come either from the charcoal or from the ore, and no doubt lowered the melting point of the slag considerably below that of silica, but the large amounts of iron (corresponding to 11% and 8% FeO respectively) lost to the slag were necessary to bring the melting point down to a practical level. Hu Boyuan (*1946*, p. 801) and Li Renkuan (*1959*, p. 199) also discuss the composition of the slags of the traditional blast furnaces of Sichuan.

舊法冶鐵爐斷立面

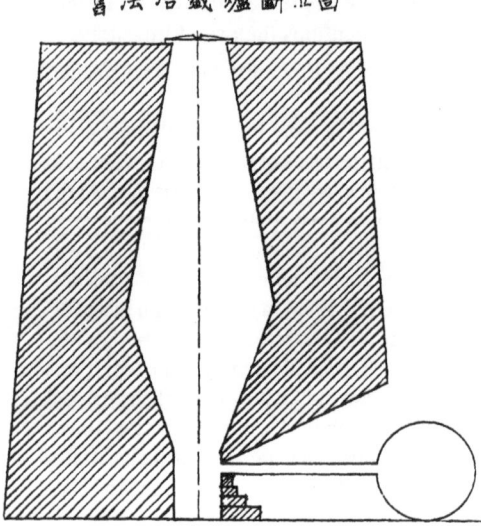

Figure 12. Vertical section through a blast furnace in Qijiang 綦江 County, Sichuan, reproduced from Luo Mian (*1936*, pp. 19–21, figs. 1–3).

Li-yün-pa[18] lies in a broad valley which we passed through to the NW, upstream on a river exploited by large and small bamboo scoop-wheels. Before us ascended a thick column of smoke produced by the blast furnace works Mu-tu-ba'rh, which we reached after a short hike. The ore which comes here for smelting is limonite [*Brauneisenstein*] from Kan-ya-dse,[19] 15 *li* [7·5 km] from the blast furnace on the road to Wan-schou-tschang.[20]

There is one blast furnace, 8 m high, with a square cross-section outside and with an outer framework of wooden poles. The taphole is located in an arch of coarse masonry, while on the opposite side is the opening through which the blast is led to the hearth. There is only one taphole, used for both iron and slag. The blast is produced in a cylindrical wooden [piston-]bellows. It is blown diagonally downward to the furnace bottom through bamboo pipes fitted with a tuyère of fireclay. The fuel and reducing agent are charcoal, half-charred wood, and fresh wood. The only flux used is slag, not limestone. The top-gas escapes to the open air through a circular mouth, 38 cm in diameter. A mortared inclined plane, over which the charge is carried in baskets, leads to the mouth of the furnace.

18. Liyinba 里隱壩, a small village (pop. 800 in 1993) in approximately the vicinity described by Cremer. See Pu Xiaorong (*1993*, p. 448).

19. Ganbazi 干壩子.

20. Wanshengchang 萬盛場, in modern Nantong Mining District 南桐礦區.

Figure 13. Photograph of a blast furnace somewhere in Sichuan, reproduced from Luo Mian (*1936*, pp. 19–21, figs. 1–3).

The furnace is tapped 10–11 times per day, producing in all 900 kg of pig iron and using 1400 kg of fuel [per day]. The workers receive 120 Marks per year each, and a similar sum is paid by the owner in tax to the government. The refractory stone for the lining of the furnace comes from Tschang-tschung-kou, 15 *li* [7·5 km] away.[21]

If this works at some time in the recent past used limestone as a flux, then the old slag from that time would have contained a fair amount of lime (CaO), and could have been of some use as a substitute for limestone; otherwise it is difficult to imagine what advantage there would have been in the continuous recirculation of slag through the furnace.[22] In the following pages we shall see several clear examples of highly developed techniques being degraded in the nineteenth and twentieth centuries under the pressure of changing

---

21. Cremer (1913, p. 51).
22. Percy (1864, p. 520) discusses a report that old slag can be used as a flux in British blast furnaces.

Box 3. Description by Luo Mian (*1936*, pp. 18–31) of a blast furnace in Sichuan in the 1930s. Cf. Figures 12–13.

The smelting furnace. The dimensions of the furnace are different in different places. Usually the height is 2 *zhang* 4 *chi* [8 m], mouth [diameter] 2 *chi* 8 *cun* [93 cm], [inner diameter of the] belly at the widest point [the boshes], 5 *chi* 8 *cun* [193 cm], bottom [diameter] 9 *cun* [30 cm].[1] The interior has the shape of a vase [*tanzi* 罎子] of the above dimensions, widest in the middle and diminishing toward the top and bottom, as in [Figure 12], which shows a section of the smelting furnace at the most recently built ironworks in Qijiang 綦江 [County]. The mouth is a round hole into which wood, charcoal, and ore are charged and from which smoke escapes. The taphole is ca. 6 *cun* [20 cm] higher than the bottom; it has diameter 8 *cun* [27 cm, surely a typographical error for 8 *fen* 分, 2·7 cm], and is used for tapping both iron and slag.

The exterior of the furnace is square. It is narrowest at the top, ca. 1 *zhang* 4 *chi* [4·7 m], and widest at the bottom, 1 *zhang* 8 *chi* [6 m]. [Figure 13] gives an overall view. The work area [the tapping arch] looks like a city gate… The place where blast is blown in [the blowing arch] is at the side or the back, and has the same form as the work area [tapping arch]. The mouth [tuyère] of the pipe from the windbox is inserted into the furnace at a height of ca. 1 *chi* [33 cm]; it has an adjustable prop, so that the height can be varied as desired. The blast is operated either by human power or by water power. The blast is cold, though some ironworks in Qijiang have recently changed over to hot blast.

The furnaces are built of refractory sandstone, which is acidic. In the interior, above and below the tuyère, it is plastered with a mixture of salt, clay, and sand to prevent the corrosion of the sandstone by basic materials. The plaster is usually composed of ca. 1 part salt, 4 parts clay, and 5 parts sand; some use *danba* 胆巴 (a magnesium salt) mixed with salt and clay.

---

1. Clearly the furnace diagrammed in Figure 11 does not have the same proportions as these dimensions.

---

economic conditions, and this works, I suggest, is likely to be another such example. Note also that water power apparently was not used for the blast here.

Mineral coal was rarely if ever used in blast furnaces in Sichuan. It was sometimes used here in copper smelting,[23] blacksmithing,[24] puddling (see below), and steelmaking,[25] and it was also used in large and small blast furnaces in Hunan,[26] but there would seem to have been severe technical problems involved in using mineral coal in the Sichuan blast furnaces. In 1940 *one* traditional ironworks in Sichuan used mineral coal as its blast furnace fuel. This was the Shujiang Ironworks 蜀江鐵廠 at Long-wangdong 龍王洞 in Jiangbei 江北 County, where it is said that the engineer Deng

---

23. Cremer (1913, p. 58).
24. Cremer (1913, p. 43).
25. Luo Mian (*1936*, p. 27).
26. von Richthofen (1877–1912, vol. 3, pp. 455–456); Lux (1912).

The sandstone is supplied seasonally by stoneworkers. The furnace as a whole can last more than 200 days, but it is never worked for more than 200 days in a year. The next year it is rebuilt. The expenditure [per year] for the furnace and tuyères is somewhat over 1,000 *yuan* 元. Assuming the worst case, a production of only 250 tonnes of pig iron per year, then the cost of the furnace is about 4–5 *yuan* per ton, which is not expensive; the disadvantage of this type of furnace lies rather in the quality of the pig iron produced.

Production of pig iron. After the bottom of the smelting furnace has been dried a layer of 400 *jin* [240 kg] of semi-charred charcoal is laid out on the bottom, followed by 600 *jin* [360 kg] of calcined ore. Further layers are added, alternating in the same way, until the top of the furnace is reached, giving a total of about 14,000 *jin* [8·4 tonnes] of ore. Then the bottom layer is ignited and the blast is started to maintain combustion. After about a day [morning to evening] the molten iron begins to flow. Outside the taphole the workers prepare a bed of fine sand which they rake to form a mould for a flat plate. The taphole is opened by striking with an iron rod, and iron and slag flow out together.

As they flow into the mould the rake is used to remove the slag; the iron left behind in the sand mould, when it has cooled, is the cast-iron plate [the pig]; it is ca. 2 *chi* 5–6 *cun* [ca. 85 cm] long, 1 *chi* 4–5 *cun* [ca. 50 cm] wide, and weighs ca. 100 *jin* [60 kg].[1] After the first day tapping can be done from time to time as convenient. The quantity of iron tapped is highly dependent on whether the 'heat' [*reli* 熱力, i.e. temperature] is sufficient; therefore it is important not to neglect the 'heat'. After the first tapping, fuel and calcined ore continue to be charged alternately as described above. As a general rule more than 4,000 *jin* [2·4 tonnes] of cast iron can be produced from 12,000–13,000 *jin* [7·2–7·8 tonnes] of calcined ore. By custom more than 5,000 *jin* is considered excellent, 4,000 *jin* normal, and 3,000 *jin* inferior. [These quantities are 3, 2·4, and 1·8 tonnes respectively, presumably per day.]

1. The thickness of the plate would then be ca. 1·8 cm.

Liangqin 鄧郎苓 only succeeded with the new fuel after more than 100 unsuccessful attempts.[27]

A more circumstantial description of blast-furnace iron-smelting in Sichuan was given by Luo Mian in the early 1930s. His illustrations are reproduced in Figures 12–13, and the essential part of his description is translated in Box 3. It will be noticed that the internal form of the furnace is quite different from that shown by Széchenyi (Figure 11). A survey in 1940 states that this more angular form was more modern (see Figure 14);[28] it is reminiscent, in fact, of some nineteenth-century British blast furnaces, and it is quite possible that its adoption was due to some Western influence. A curiosity is that this furnace does not have the wooden frame mentioned in the earlier descriptions.

◆ ◆ ◆

27. Wang Ziyou (*1940*, p. 3).

28. Note also the somewhat similar diagram of older and newer blast furnaces in Sichuan given by Hu Boyuan (*1946*, p. 800).

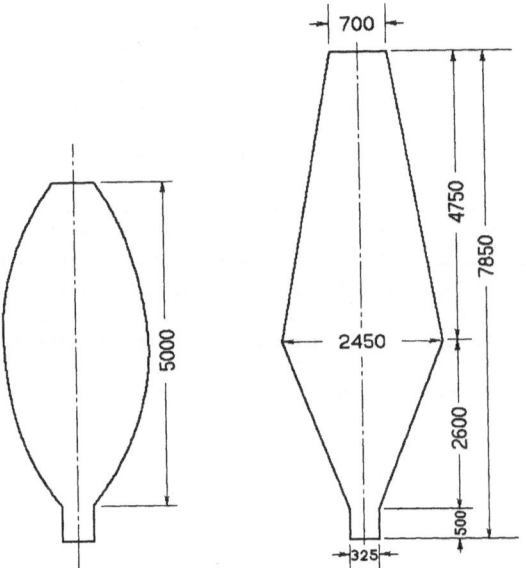

Figure 14. Internal form of two traditional blast furnaces in Sichuan in 1940, redrawn from Wang Ziyou (*1940*, foldout fig. 1). Dimensions are given in mm. On the left is the 'older' form, on the right the 'newer' form. Cf. Figures 11, 12.

A survey published in 1940 gives chemical analyses of samples of cast iron produced in three counties in Sichuan.[29] The sulphur content of this cast iron is low because the ore was carefully calcined to remove sulphides, and

---

29. Wang Ziyou (*1940*, p. 5, table 11). Analyses of eight pig iron samples published by Luo Mian (*1936*, p. 22, tables 6–7) are very different from these analyses, and in fact seem so bizarre that one must suspect an error on the part of the analyst: the reported silicon contents approach that of modern pig iron, the sulphur contents are so high as to make the iron useless for many purposes, and the phosphorus contents are also high in comparison with the analyses given by Wang Ziyou. Several observers mention poor separation between iron and slag, due to excessive slag viscosity, and the cast-iron plates from the blast furnace are described as 'slaggy'. A chemist asked to analyse such a physical admixture of very different materials has a choice to make, according to the purpose of the analysis: he can separate out the metallic phase by physical means (e.g. by melting it) and analyse that alone, or he can use a variety of means to bring the entire sample into solution so that the analysis will apply to the heterogeneous sample as a whole. The first approach would be more appropriate in an analysis of pig iron, but it would seem that the chemists who did the analyses reported by Luo Mian (perhaps more accustomed to analysing ores than metals) took the second approach, and the analyses include the slag inclusions. These would contain a large amount of silica (SiO_2), and might contain a fair amount of sulphur if calcining was imperfect but a limestone flux was used in the blast furnace. Some curious analyses are also reported by Hu Boyuan (*1946*, p. 801).

because charcoal has very low sulphur. The high chemical activity of charcoal means that it burns at a lower temperature than mineral coal or coke, and this means that not much silicon is reduced and enters the iron. It is surprising to see the very low carbon contents in these analyses: we should expect the carbon content of pig iron from a blast furnace to be between 3·5 and 5 per cent. The liquidus temperature of the sample from Qijiang would be quite high, about 1380° C.

County	C %	Si %	S %	P %	Mn %
Qijiang 綦江	2·10	0·20	0·05	0·70	0·05
Weiyuan 威遠	3·00	0·18	0·06	0·22	0·25
Fuling 涪陵	3·30	0·19	0·05	0·22	0·01

The puddling furnace

The cast iron produced in the blast furnace was converted to wrought iron in what I shall call a *puddling furnace*. In Chinese it was called a *chaolu* 炒爐; the same word was used for the *fining hearth* described earlier, but as we shall see further below, the operation of the Sichuan *chaolu* had many characteristics in common with the 'puddling' process patented by Henry Cort in 1784.

Cremer observed this process in Sichuan in 1905, and he was the first to refer to it as 'puddling' (*Puddeln*).[30] The furnace and its operation were described in detail by Luo Mian in 1936.[31] He gives the diagram reproduced here as Figure 15. He also gives two photographs which, unfortunately, are so unclear that reproducing them here would be pointless. The furnace is built entirely of a type of refractory sandstone which is common in Sichuan.[32] Charcoal is burned in the closed firebox. Blast is blown into the firebox, and the flame proceeds downward into the puddling bed.

> *Operation.* [See Figure 15.] First semi-charred charcoal [*chaitan* 柴炭] is burned in the stone firebox **a** to heat the puddling bed. Then broken pieces of pig iron are charged into the puddling bed **b** and the blast is increased, blowing through **c** to **a**. The flame passes through the stone aperture **d** and heats the cast iron thoroughly. The flame gradually becomes the colour of mung beans; [the iron] is stirred [puddled] with a wooden pole; then the colour [of the iron] turns from red to white, the

---

30. Cremer (1913, p. 51). Earlier von Richthofen (1907, p. 501) referred to *Puddelöfen* in parts of Shanxi, but gave no details.

31. Luo Mian (*1936*, pp. 23–26).

32. Yang Kuan (*1960*, p. 187) calls this *bai paoshi* 白泡石, 'white afrodite'.

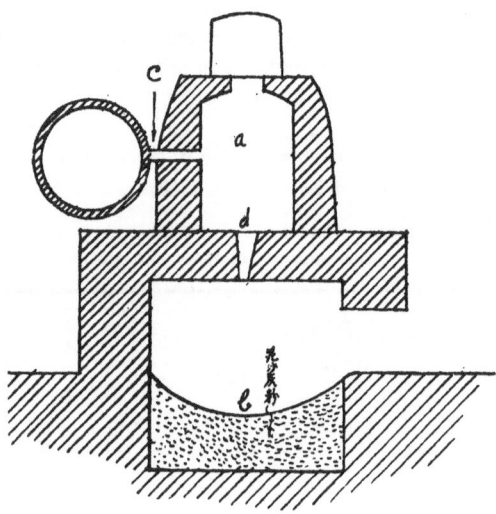

Figure 15. Diagram of a puddling hearth in Sichuan, reproduced from Luo Mian (*1936*, p. 24, fig. 4). **a.** Firebox, height 2 *chi* 尺 8 *cun* 寸 (93 cm). **b.** Puddling bed, diameter 2 *chi* (67 cm). **c.** Blast pipe. **d.** Flame channel.

temperature being about 1000° C. It is puddled again with a wooden pole until the iron breaks up into a granular form. Then a small amount of ironsand (iron oxide, commonly called *hongzi* 紅子 ['red stuff']) is added while the blast is worked and stirring continues. [The iron] gradually melts, and as puddling continues it goes from free-flowing [*xi* 稀] to 'dry' [*gan* 乾, i.e. viscous] and transforms to a plastic state. At this time the blast is reduced and [the iron] is formed into a ball; this is wrought iron [*shutie* 熟鐵], commonly called *maotie* 毛鐵 ['semi-finished iron']. The entire operation takes about 20 minutes.

The semi-finished iron is then heated in an ordinary smithy hearth and hammered to force out the slag which it contains. From 100 *jin* 斤 of semi-finished iron about 75 *jin* of wrought iron can be obtained. This is then hammered into plates, bars, rods, and the like and sold in the markets.

... It is customary to pay the puddler a combined price for his labour and the charcoal which he uses. For the puddling of 100 *jin* 斤 [60 kg] of pig iron to semi-finished iron the fixed price for labour and charcoal is 3 *jiao* 角 [0·3 *yuan* 元; the pig iron itself cost 5 *yuan*]. The labourer is not permitted an allowance for lost metal [*huohao* 火耗], but he is permitted to add as much ironsand as he wishes; therefore for every 100 *jin* of pig iron given to the puddler the return is about 100 *jin* of [semi-finished] wrought iron. However, when this is hammered in the smithy hearth, although the price of labour and charcoal is the same, 3 *jiao*, so much slag is forced out of the iron that the allowance for lost metal is about 25 per cent: if the smith is given 100 *jin* of semi-finished iron the return is only 75 *jin*.

As the iron is melted the carbon in it is oxidised by the oxidising flame together with a slag rich in FeO. The slag is formed partly by the addition of red ironsand (presumably a mixture of quartz and hematite, SiO_2 and Fe_2O_3), partly by the combustion of some 25 per cent of the pig iron charged. Unfortunately Luo Mian gives very little quantitative data on this process, and I have not found any elsewhere, but the small size of the puddling bed and the short time required for the puddling operation suggest that the amount of pig iron charged was perhaps 10–20 kg. The amount of fuel needed probably varied greatly with the skill of the puddler: this would be the reason for requiring him to supply the fuel himself.

In the puddling process, as the carbon content of the iron decreases, its melting point (liquidus) rises. At the same time the temperature in the furnace also rises, probably coming close to 1500° C. Therefore the iron in the puddling bed is through most of the process in the form of a paste composed of microscopic crystals of lower-carbon iron in a higher-carbon liquid. In the neighbourhood of 1500° C the liquid has about 0·5 per cent carbon and the solid about 0·1 per cent.[33] As the carbon content continues to decrease, more low-carbon iron precipitates from the melt, the proportion of liquid decreases, and finally the plastic state mentioned in the description above is reached, at which time the iron has been nearly fully decarburised. English puddlers spoke of the iron 'coming to nature', and in the large puddling furnaces of the nineteenth century this was quite a spectacular phenomenon.[34]

In the 'fining' process used in the Dabieshan region, described above, the fuel was mixed directly with the iron. There the temperature appears to have been lower, and the decarburisation presumably took place with the iron in the solid state. Decarburisation of iron in the solid state is much slower than in the liquid state, and this process, though fuel-efficient compared with early Western fining processes, was probably less fuel-efficient than the puddling methods described above. Its great advantage was no doubt the use of a lower temperature, which meant that local materials, less refractory than the sandstone available in Sichuan, could be used for the furnace.

The 'puddling' process patented in Britain by Henry Cort in 1784 was

---

33. Details of the iron–carbon phase system are given by Hansen (1958, pp. 353–365). The details referred to here are in his fig. 203, p. 356.
34. Percy (1864, p. 656). Gale (1977, figs. 38–43) gives some marvellous photographs of the English puddling process as it was still being practised in the 1950s.

in principle very like the Sichuan process described above.[35] Mineral coal was burned in a firebox and its flame was used to heat cast iron in a separate 'puddling bed'. The puddler performed essentially the same operations as described here; 'to puddle' is a nearly-obsolete English word meaning 'to stir about'. There were two essential differences: the British process used mineral coal as the fuel,[36] and it was a much larger-scale operation than the Sichuan process. The puddling bed was many times larger, tonnes of iron per day were converted to wrought iron, and puddling was 'probably the severest kind of labour in the world.'[37] It is unlikely that the work of the Sichuan puddlers was equally severe, but it cannot have been as effortless as the above description might suggest.

It is a surprise to see that mineral coal was rarely used in puddling in Sichuan. The great benefit of Henry Cort's innovation was that it separated the fuel from the iron and therefore made possible the use of coal in converting cast iron to wrought iron. European fining processes consumed prodigious amounts of charcoal,[38] and mineral coal could not be substituted directly because its high sulphur content was in large part taken up by the iron. Chinese fining processes were much more fuel-efficient, and the separation of the fuel from the iron in the Sichuan puddling process no doubt increased this efficiency to some extent; perhaps, then, a need for a cheaper fuel was rarely felt.

Two types of puddling furnace in use in Sichuan which did use mineral coal are described briefly by Yang Kuan.[39] They were used in the Great Leap Forward period, and presumably also earlier. He gives the diagram redrawn in Figure 16. It can be seen that these furnace designs provide much greater separation between fuel and iron than the one discussed above (Figure 15).

---

35. On this process the best general technical discussion seems still to be that of John Percy (1864, pp. 627ff). A few other useful references, among many, are Turner (1895, pp. 281–314) (detailed description and technical explanation); Rosenholtz & Oesterle (1938, pp. 89–96); Gale (1977, figs. 38–43) (marvellous photographs); Mott & Singer (1983) (the history of the invention); Paulinyi (1987) (a broader historical study).

36. Though wood was sometimes used when the process was adopted in heavily forested regions such as northern Sweden and the Carinthian Alps. Percy (1864, pp. 686–688).

37. Percy, himself a physician, wrote in his *Metallurgy*: 'Most puddlers work until 50 years of age, and many even afterwards. Puddling is probably the severest kind of labour in the world; yet many puddlers attain the ripe age of 70 years, or more. The majority die between the ages of 45 and 50 years; and, according to the returns of medical men to the Registrar, pneumonia, or inflammation of the lungs, is the most frequent cause of their death. This is what might have been anticipated from the fact of their exposure to great alternations of temperature under the conditions of physical exhaustion. Mr. Field, optician, Birmingham, informs me that puddlers are moreover liable to cataract. induced by the bright light of the furnace; that he has seen a great number of such cases, and supplied the patients with glasses.' (1864, pp. 656–657)

38. See e.g. Percy (1864, pp. 596, 601, 607, 615).

39. Yang Kuan (*1960*, pp. 187–188).

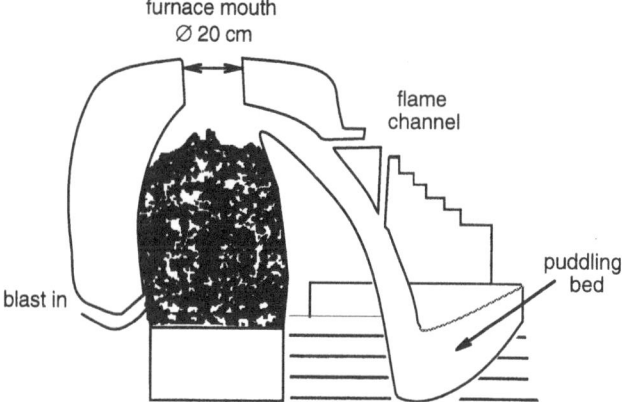

Figure 16. Puddling furnace used in Sichuan in 1958, redrawn from Yang Kuan (*1960*, p. 187, fig. 69). The dimensions are not given. Mineral coal or charcoal is burned in the large chamber; blast is blown in at the left. The flame travels through the long passage to the puddling bed at the right.

The iron industry of Sichuan

There was a major iron industry in Sichuan in very early times.[40] Under the state monopoly of the Han period a number of 'Offices for Iron' (*tie guan* 鐵官) produced iron in large blast furnaces. In the Song period, according to the researches of Robert Hartwell,[41] production in Sichuan in the year AD 1078 may have been on the order of 10,000 tonnes for a population of about 12 million. If we were to suppose for the sake of argument that *per capita* production was of the same order of magnitude whenever there was peace and prosperity, we might arrive at an annual production as high as 20,000 tonnes in the early nineteenth century.[42] This production might have required 100 blast furnaces. Obviously the highly speculative reasoning I give here is not of great value, but it shows the sort of scale which the iron industry may have had in recent centuries.

Virtually nothing is available in the way of economic statistics for Sichuan in the nineteenth and early twentieth century.[43] We do not know

40. See e.g. Wagner (1993, pp. 250–253).

41. Hartwell (1963, p. 56).

42. Skinner (1987) has shown that the reported population statistics for Sichuan in the nineteenth century are the result of systematic falsification by provincial clerks. Careful analysis of the methods of falsification lead to the result that a population of 22 million in 1813 is the result of a reliable census. Other reasoning gives the provincial population in 1833 as 25.4 million and in 1853 as 28.5 million. Cf. the much higher estimates of Bielenstein (1987, p. 118).

43. Cf. Bramall (1993).

how many ironworks there were or, with any degree of reliability, how much iron they produced.[44] We may guess that China's generally unhappy conditions in this period were felt in Sichuan, and that its industries suffered to some degree. On the other hand, Sichuan's isolation probably protected the iron industry from the worst shocks. Production had always been very largely aimed at local demand. This demand may have decreased because of bad times, but competition from cheap foreign iron would not have been severe before steamship transport through the Yangzi Gorges became a commercial reality in the 1920s.[45]

Though we know something of the economic history of iron production in Sichuan, we know nothing of its technology before the late nineteenth-century descriptions discussed above. We see it here in a time of change. Two important features of blast furnace operation seem to have been going out of use: limestone flux and water-powered blast. In addition, it seems that the technology may have been modified in some ways because of foreign influence.

The use of a limestone flux seems to have disappeared almost completely by 1900. The primary functions of the limestone flux were to reduce the loss of iron to the slag, to lower the melting point of the slag, and to reduce the sulphur content of the pig iron produced. It would seem from the analyses reported above that sulphur in iron was not a major problem here. The pig iron is several times described as 'slaggy', because the high melting point of the slag meant poor separation of slag and iron. This was probably not a major problem, since the pig iron was to be melted again in the foundry or puddling furnace, where better separation could be effected without great trouble. The cost of limestone may have risen in Sichuan at the end of the nineteenth century as brick-and-mortar building became more common and cement production expanded: in Cremer's travels in Sichuan in 1905 he noted numerous lime kilns wherever he went.[46] But besides the cost of limestone another major disadvantage of its use is that a basic slag attacks the stone lining of the blast furnace, reducing its useful life. In the generally poor economic conditions of the time, operating blast furnaces without flux was a sensible way of cutting costs.

---

44. Zhang Xiaomei (*1939*, p. Q1) estimates annual iron production by traditional methods in Sichuan at 21,500 tonnes, but does not tell how this estimate was arrived at.

45. A Japanese steamship had negotiated the Gorges as early as 1895, and English attempts succeeded in 1898. In the period 1911–21 three American and four British steamships plied the Yangzi as far as Chongqing. The real impact of steamship transport was not felt here, however, until Chinese firms came into the trade in the 1920s. Jiang Tianfeng (*1992*, pp. 224–5); cf. Dautremer (1911, pp. 6–7); Rawski (1989, pp. 44, 191ff).

46. Cremer (1913, *passim*).

Water power continued in use rather longer, perhaps until around 1920,[47] but later all reports indicate that human labour was used for the blast. Perhaps some electrical or steam-powered blowing machines were also in use, but the overall impression is that of a transition from water power to labour power. This presumably reflects an economic development in which the price of capital rose while the cost of labour fell. Besides the obvious direct saving of capital for the building of water-wheels, the transition to human labour also saved capital in another way. Water power could obviously be used only in certain locations. A labour-powered blast furnace, not being restricted in this way, could be located according to other considerations, such as land prices or proximity to raw materials, markets, or transportation, and some saving in capital costs, operating costs, or both would have resulted.

In the 1930s there were also changes in the technology of Sichuan's iron industry which were due to foreign influence. The 1936 survey describes a cementation steelmaking furnace in Weiyuan County 威遠縣 which clearly is of foreign design.[48] The cementation steelmaking was a German invention of the late sixteenth century, and developed into the most important steelmaking process of the early nineteenth century. The introduction of the Bessemer process provided cheaper means of making steel, and by 1900 the cementation process was essentially obsolete, though it continued in use in Sheffield and a few other places until as late as the 1950s.[49] It is a surprise and a puzzle to find this process in use in Sichuan in the 1930s. When was it introduced, and by whom?

Another surprise has already been mentioned, the distinction between 'older' and 'newer' internal shapes of blast furnaces mentioned in a report of 1940.[50] The 'newer' form resembles that of some nineteenth-century European blast furnaces, curiously enough a form which in the mid-nineteenth century was being replaced by more rounded forms like the 'older' form in Sichuan.[51] It is perhaps less certain that this 'newer' form in Sichuan was introduced from the West; if it was it is one of a very few examples of changes being made to traditional Chinese iron-production technologies through foreign influence. The usual pattern of foreign influence involved the total replacement of existing technologies with foreign technologies, usually under foreign control.[52] Again: when was this influence, and through whom?

---

47. Notices of water power used in blast furnace operation in Sichuan include: Széchenyi (1893, pp. 678–679) (see Box 2); DuClos (1898, p. 313); Cremer (1913, p. 58); Way (1916, p. 22).
48. Luo Mian (*1936*, pp. 35–38).
49. Barraclough (1976).
50. See Figure 14.
51. Percy (1864, pp. 350ff, esp. pp. 441, 475–491, 559).
52. See e.g. Brown (1978; 1979).

The first two technological developments listed above – concerning the flux and the blast – would seem to be the cumulative result of decisions made by numerous entrepreneurs adapting to changing economic conditions. In one sense these developments represent a technological degeneration, since they involve the abandonment of previous innovations; however in another, more detached, sense they may be seen as progressive, for they make for a more efficient iron industry in the specific context in which it must function. In dealing with technological change it is always well to remind oneself of these two sides of the question of progress.

The latter two developments – concerning possible foreign influence – probably involved some form of government intervention. From the 1911 revolution to 1937, Sichuan was controlled by a number of mutually hostile 'warlords' (junfa 軍閥, regional militarist rulers).[53] After the 'War of the Two Lius' (Liu Wenhui 劉文輝 and Liu Xiang 劉湘)[54] of 1932, the most important iron-producing areas of the province were under the control of Liu Xiang. Though in general as bellicose and mediocre as any of the others, Liu Xiang did do a certain amount toward industrial development, and research into his 'Min Shêng Industrial Company' and 'West China Development Corporation' may well show that they were the mediators of foreign influence in the traditional iron industry of Sichuan.[55]

In 1937, with the start of full-scale hostilities with Japan, Sichuan took on a new importance. Chiang Kai-shek's Nanjing government moved here, as did the greater part of China's universities and scientific research institutes.[56] Attempts to build modern steelworks here failed, and the traditional iron industry was now the only provider of iron and steel. Numerous studies were made of ways to improve the traditional technology, and it would seem that some at least of these were successful.[57]

After the Second World War the traditional iron industry seems to have continued in full-scale operation, and efforts to develop its technology continued. For example in the early 1950s experimental work commenced in

---

53. See especially Kapp (1973).

54. Boorman (1967–71, vol. 2, pp. 395–398, 417–419).

55. Kapp (1973, pp. 58, 152 n. 85–86) cites some sources on these enterprises. Note also Luo Mian (*1936*, pp. 5–7).

56. During the war Joseph and Dorothy Needham were in Sichuan with the Sino–British Science Co-operation Office, and their *Science outpost* (1948) provides a wonderful glimpse of the intense atmosphere in the by this time almost totally isolated province.

57. E.g. Wang Ziyou (*1940*); Zhu Yulun (*1940*); and many other articles in the journal *Kuangye banyue kan* 冶半月刊 (Mining and metallurgy semimonthly) in the war years. Hu Boyuan (*1946*, pp. 810–811) gives information in tabular form on many of the ironworks established in Sichuan in this period. Meanwhile, in the Shaanxi–Gan-su–Ningxia Border Region Soviet, research was also being carried out on the im-provement of the traditional iron-production techniques of this region. Mao Zedong (1980, p. 160).

Hechuan 合川 and Jiangbei 江北 Counties on ways to prolong the operating life of the Sichuan blast furnaces.[58] This type of research was no doubt of great value in the Great Leap Forward in Sichuan, though I have not found much information on it.[59]

---

58. Li Renkuan (*1959*). As noted above (Box 3), a normal furnace campaign was on the order of 200 days before repairs became necessary. A blast furnace in Jiangbei operated continuously from November 1954 to August 1958; one in Hechuan was blown in during July 1955 and was still in blast at the beginning of 1959.

59. But note Goodman (1986, pp. 91ff).

6

Crucible smelting in Shanxi

Shanxi seems fitted out by nature for the iron industry, with the world's largest deposit of coal, reasonably large reserves of iron ore and limestone, and not very much else in the way of raw materials for industry.[1] Coal mining and iron production were sideline occupations for a large part of the peasant population, but there seem also to have been large areas in which iron-making was the *only* occupation. In Yincheng 蔭城, for example, a town with a population of perhaps 5,000, people told a visitor in 1898, 'We eat iron'.[2]

Crucible smelting

The method used here for smelting iron was very different from what we have seen above. A mixture of crushed iron ore and coal was packed in crucibles, and the crucibles heated in a stall furnace with more coal. The coal in the crucible reduced the iron oxides in the ore.

The earliest published description of the method seems to be that of von Richthofen in Dayang 大陽, Shanxi, in 1870. It is worth translating in full for the light it also sheds on the general conditions of the iron industry in Shanxi. No illustrations accompany this description, but Figures 17–19, taken from later publications, provide some help.

> Meeting innumerable animals and coolies on the pack road carrying anthracite, one expects to find a large-scale mine; but both coal mining and iron manufacture in this region have the character of all Chinese industry: rough, exceptionally diminutive, and nevertheless of an extraordinary perfection. One is astounded, arriving at these much-discussed places, to see merely hundreds of small establishments among which the work is distributed. One finds nothing which even remotely resembles a European blast furnace.

---

1. Background material for Shanxi's economic geography includes Williamson (1870, pp. 151–169, 287–363); von Richthofen (1872, pp. 27ff, 94ff; 1877–1912); Nyström (1910); Nyström (1912); Corbin (1913); Anon. (*1920*); Yang Dajin (*1938*, pp. 341–348); Qiao Zhiqiang (*1978*); Anon. (*1985a*).
2. Shockley (1904, p. 850).

Figure 17. Stall furnace for crucible iron smelting in southern Shanxi, photograph reproduced from Kocher (1921, p. 10, fig. 1).

Figure 18. Lumps of cast iron and puddled wrought iron in southern Shanxi, photograph reproduced from Kocher (1921, p. 10, fig. 3).

The iron smelter is situated on a slightly inclined floor, $2^{1/2}$ m long and $1^{1/2}$ m wide. On the two long sides are walls, $1^{1/4}$ m high; the third side, towards which the floor ascends, is open; and on the fourth is a small and primitive hut for the bellows and two people who work it. The floor is covered with small pieces of anthracite, the size of a fist. On this are placed about 150 crucibles of refractory clay, 15 inches high [38 cm] and 6 inches wide [15 cm], which are filled with a mixture of small pieces

Figure 19. Crucible smelting of iron in progress in Gaoping 高平 County, Shanxi, photograph reproduced from Shockley (1904, fig. 1, facing p. 854).

of anthracite and crushed iron ore.[3] All the spaces between crucibles are carefully filled out with anthracite, and a layer of the fuel is spread on top. Sometimes a second layer of 150 crucibles is laid over the first. Over this is laid more anthracite and on top a layer of shards of old crucibles. The whole heap is ignited, and air is blown in. When everything is burning and the heat is great, the blowing is stopped, since the natural draught is sufficient to maintain the heat.

If the intention is to make cast iron [*Roheisen*], the crucibles are taken out after a certain period of time and the contents cast as flat plates; the result appears to be a clean white steelmaking pig iron.[4] If wrought iron is desired, the heap is allowed to burn out and cool off over a period of four days. The crucibles are then taken out and broken. In this case the iron is in the form of a hemisphere.

These two types of iron serve as the raw material for a wide variety of manufactures. Their further treatment of one sort or another for particular purposes is kept secret by the individual factories, and some of these have acquired a great reputation for the preparation of kettles, ploughs, or other equipment.

3. It seems certain that von Richthofen meant English inches when he used the word *Zoll*. The text gives 5 inches, which is very unlikely to be correct. The version in von Richthofen (1872, p. 30) gives 15 inches. Even with this correction these are the smallest crucibles reported anywhere in Shanxi.

4. *Stahleisen*, a pig iron suitable for open-hearth steelmaking, with low phosphorus and sulphur and fairly low silicon; cf. Anon. (1939, p. 782).

A third type of raw iron is also prepared by casting the molten metal in water to form drops. This type is added in various quantities to the other types in order to suit various purposes.

The best product is the wrought iron, which is far superior to that of Europe and possesses great malleability. The Chinese also excel in the casting of very thin objects, such as the iron pans [woks] used for cooking; this is an art which they understand everywhere, but Shanxi is its home.

It is of great interest to go around to the different establishments and see everywhere these simple methods used which have served since ancient times. It is clear that this great perfection must be ascribed not only to experience but also to the quality of the raw materials. Everything they need is supplied by the strata of productive coal formations which are only a few hundred feet thick. Of the very widespread iron ores only the purest and most easily smelted are used. Clay and refractory material are also found in great quantities. But the most important material is anthracite.[5]

Numerous other sources also exist for the crucible smelting process as it was practised in Shanxi and elsewhere, and we find considerable variation in the details from place to place.[6] Typically the crucibles might be 15–20 cm in diameter and 50–100 cm high; the charge in each, 15–25 kg ore and 4–6 kg coal; the number of crucibles in the furnace from under 100 to over 300; the heating time 1–3 days; and the yield of iron from ore 20–40 per cent.[7] Natural draught alone was sometimes used, but more often, as here, a man-powered blast was used during part or all of the process. The iron produced in this way was normally in the form of a very slaggy bloom, with a carbon content in the range 1–3 per cent. This was either decarburised by any of a number of processes (e.g. Figure 20) to make wrought iron, or carburised in a cupola or crucible furnace to make cast iron.

◆ ◆ ◆

---

5. von Richthofen (1907, pp. 498–499); cf. von Richthofen (1872, p. 30).

6. The most important sources seem to be the following. In Western languages: Davidov (1872a; 1872b); von Richthofen (1872, pp. 30, 34; 1877–1912; 1907, pp. 498ff); Henderson (1872, 74–7, 85–6, 119–20); Henderson in Day (1875, Appendix, pp. 29–32); Shockley (1904); Nyström (1910, p. 398); Anon. (1910); Read (1921); Kocher (1921); Licent (1924, vol. 2, pp. 623–6); Tegengren (1923–24, pp. 323–7); Foster (1926); Wang Jingzun & Wang Yuelun (*1930*, Eng. pp. 109–112, Ch. pp. 85–87); Dickmann (1932); Hara (1992). In Chinese: Yang Dajin (*1938*, pp. 344–5); Ding Wenjiang (*1956*, pp. 369–72); Kong Lingtan (*1957*); Yang Kuan (*1960*, pp. 95–99); Fan Baisheng (*1985*). In Japanese: Anon. (*1920*, pp. 596–7); Hara Zenshirō (*1993*). In compiling these sources I have benefited greatly from a correspondence some years ago with Dr. Bennet Bronson.

7. See e.g. the comparative table given by Tegengren (1923–24, p. 326).

Figure 20. Operation of a furnace in Shanxi for converting cast iron to wrought iron, photograph reproduced from Alley (1961a).

As von Richthofen observed, the scale of enterprise in Shanxi in the nineteenth and twentieth centuries was very small. (He was not correct in supposing that the same was true all over China without exception, as will be seen below in the discussion of Guangdong.) We may perhaps explain this by noting the tiny capital cost of a crucible iron smelter: 'A capital of thirty dollars would be sufficient to start this complete set of foundry and iron works, including the purchase of a stock of coal, ore and fire-clay. These substances are all close at hand, and cost only a trifle.'[8] It seems that almost anyone could start an ironworks, and that the technology used did not provide any great economies of scale.

Tegengren gives some economic data on the process as he observed it, also in Dayang, around 1920 (see opposite).[9] We may note the prodigious consumption of fuel here, almost five times the weight of the iron produced: its cost represented more than half of the total cost, and a small reduction in coal consumption could have increased profits dramatically. We may wonder whether Tegengren's figures are typical for all crucible smelting. The cost of the large amounts of refractory clay needed for the crucibles would have been prohibitive in many parts of China, but in Shanxi it seems to have been

---

8.   von Richthofen (1872, p. 34).

9.   Tegengren (1923–24, p. 327). 'Catty' was formerly a commonly used translation for *chin* 斤, a measure of weight equal to ca. 0·6 kg.

almost insignificant. The clay was available in great quantity out of the same geological strata as the ore and the coal. A final point to be noted is that the fragmentation of the iron industry into small competing enterprises had driven profits down to only 4·8 per cent on the price of the product.

Expenditure [for one heat]

Raw materials:

Anthracite	2000 catties	@ 1·3 cash	2600 cash
Ore	1500 catties	@ 0·6 cash	900 cash
Hei-T'u[10] & dust coal	400? catties	@ 1·0 cash	400? cash
Clay for 64 crucibles @ 5 kilos each	320 kilos	@ 5 cash	160 cash

Labour:

Wages for 500 catties of raw iron	@ 1·2 cash per catty	160 cash
Addition for sundry expenses, 2% of the total		100 cash
	Total	4760 cash

Receipts

Sales at the smelter of 500 catties of raw iron	@ 10 cash per catty	5000 cash
	Approximate profit	240 cash

The iron industry of Shanxi

Nature's greatest gift to Shanxi was its enormous coal reserve, but poor transportation made the export of coal uncompetitive,[11] and the province's most important export to other provinces was iron. From early times iron speciality products seem to have been the leading exports: the Tang poet Du Fu 杜甫 mentioned the famous scissors of Shanxi,[12] and for centuries a large proportion of the needles used in China came from here.

This speciality trade was hit very hard by foreign competition, as von Richthofen noted:

> The competition with foreign trade is another cause of the decadence of the wealth of Shansi. If we commence with the trifling article of nee-dles, their manufacture in Shansi has almost been annihilated, by the importation of the much better and cheaper foreign article. The same

---

10. *Hei tu* 黑土, see below.

11. In 1870, 'I repeat, that coal, which costs in Shansi thirteen cents per ton at the mine, rises to four taels at a distance of thirty miles, and to over seven taels at a distance of sixty miles; also that, at Nan-yang-fu [南陽府] (Honan), coal from Hunan is used which has travelled eight hundred miles by water, and is sold at the same price with the coal mined at a distance of thirty miles from the city, but which is transported by land.' von Richthofen (1872, p. 37).

12. Liu Jixian et al. (*1982*, p. 9).

will be true, before long, in regards to guns and steel ware; and there can
be no doubt that the injurious effects of foreign competition have been
seriously felt by the iron trade of Shansi in general. Being the only note-
worthy article of export from that province, the diminished sales and reduced prices contribute to impoverish the inhabitants.[13]

He estimated the iron production of the entire province to be very roughly
160,000 tonnes per year. When Shockley visited Shanxi 28 years later, in
1898, he arrived at a rough estimate somewhat in excess of 50,000 tonnes per
year, and went on:

> When von Richthofen was in Shansi, he estimated the production of
> iron at 160,000 tons per annum, which was considered an absurdly large
> estimate by critics who had never been in the province,[14] but I have no
> doubt he was well within the truth. The district magistrate at Tsê Chou[15]
> said that the iron made in that district now was only one-fourth of what
> it was thirty years ago, which was about the time that von Richthofen
> visited the province (1870–72). If the iron-trade has declined as much in
> the rest of the province as it has here, my estimate and von Richthofen's
> would not be so very different.[16]

The effect of the shortage of iron during World War I is perhaps seen in the
estimate cited by Yang Kuan for 1916 of 70,000 tonnes per year for the
whole province.[17] An estimate of 68,600 tonnes per year for the early 1920s
is given by Wang Zhuquan 王竹泉.[18]

The coming of the railroads improved the chances of the iron industry
in some parts of the province. In Pingding County 平定縣, in 1870, von
Richthofen estimated an iron production of about 54,000 tonnes per year. In
1898 Shockley's estimate was only 18,000 tonnes, and in the early 1920s
Wang Zhuquan's estimate was 20,000 tonnes. By 1928 production in this
county may have doubled rather suddenly, though there does not appear to
have been much, if any, speciality production:

> *Annual pig iron production of Pingding County by traditional methods.*
> The Office of Public Finance[19] of Pingding County estimates that, in

---

13. von Richthofen (1872, p. 38).
14. See Henderson (1872, pp. 155–157) and the reply by von Richthofen (1872, p. 148).
15. Zezhou 澤州, modern Jincheng 晉城, Shanxi.
16. Shockley (1904, p. 871). It should be noted that F. R. Tegengren (1923–24, p. 320),
    after citing this evaluation, gives a careful criticism of von Richthofen's estimates and
    suggests that the true annual iron production of Shanxi in 1870 may have been closer
    to 125,000–130,000 tonnes.
17. Yang Kuan (*1960*, p. 95); note also Tegengren (1923–24, pp. 320–321).
18. In a report reprinted and translated in Tegengren (1923–24, Ch. pp. 305–313, Eng. pp.
    435–443). This production estimate is given on Eng. p. 321.
19. *Gongkuan ju* 公款局.

times when transportation is in order, the pig iron exported on the Zheng–Tai 正太 [Shijiazhuang–Taiyuan] Railway amounts to about 1,500 carloads per year. Assuming 20 tonnes per carload, this gives 30,000 tonnes. In addition more than 5,000 tonnes is either melted and marketed locally or transported by mule. Thus in times when transportation is in order production is around 40,000 tonnes per year.[20]

In addition to iron produced by traditional methods, a modern ironworks, established in Pingding in 1926, 'when in good running order', was producing about 500 tonnes per month.[21] Thus the local modern sector was not yet, at this time, a serious competitor of the traditional sector.[22]

In 1870 our earliest witness, von Richthofen, gave great praise to the iron produced in Shanxi, and it is odd to find that later its quality was usually reported as being quite poor. In 1911 T. T. Read published analyses of samples of iron from twelve different ironworks in Pingding: the sulphur content ranges from 0.13 to 0.61 per cent, and even the lowest of these figures is far higher than is desirable for most uses of iron.[23] By the time of the Great Leap Forward the process was considered unusable because of the high sulphur content of the product.[24] I think it is safe to say that it would be almost impossible to make needles of this iron, and the same is likely to be true for scissors and other fine wrought products. It is likely that the quality of Shanxi iron had been much better in earlier times, but deteriorated as prices fell and it became necessary at the ironworks to reduce costs drastically.

It is not easy to know what exactly the differences may have been between the crucible smelting process as observed in the twentieth century and the earlier higher-quality process which I have posited here. One possibility, however, can be seen from experiments with essentially the same process in 1908 in Höganäs, Sweden: these showed that the sulphur content of the iron produced could be reduced to 0.01–0.03 per cent by the addition of a small amount of limestone ($CaCO_3$) to the crucible charge combined with careful temperature control at about 1200° C.[25] Limestone is available

---

20. Wang Jingzun & Wang Yuelun (*1930*, Ch. p. 86; cf. Eng. p. 112). I do not know how to explain the curious arithmetic of this passage.

21. Wang Jingzun & Wang Yuelun (*1930*, Eng. p. 112).

22. In the year 1950 Pingding produced only 2,049 tonnes of iron. Anon. (*1954*, p. 81).

23. Read (1911, p. 27). Nyström (1910, p. 398) also mentions the high sulphur content of Shanxi wrought iron. Note, however, an analysis done in 1915 which showed only 0·078 per cent sulphur in a sample of Shanxi wrought iron (Tegengren (1923–24, p. 329)).

24. Yang Kuan (*1960*, p. 99).

25. Sieurin (1911, pp. 458–459). On the Höganäs process, see also Anon. (1979, pp. 316–325).

in large quantities in Shanxi,[26] and has been used in iron production in China at least since the Han period.

Another possibility is the likely role played by a material known as 'black earth' (*hei tu* 黑土), which several observers were told was essential in the crucible charge in the Shanxi process.[27] This is a kind of decomposed coal produced by the weathering of the upper strata of the coal seams; an analysis given by Tegengren indicates that it has very low sulphur (0.21 per cent) and very high ash (32 per cent) as compared with ordinary coal. It contains about 9 per cent lime (CaO), and therefore could be expected, in sufficient quantities, to be effective in removing sulphur from the iron.

The numerical data which I have compiled here for iron production in Shanxi are the best we have for any region of China in the early twentieth century, though they are clearly less reliable than the precise numbers given might suggest. Production fell drastically in the second half of the nineteenth century, and this fall hit especially hard in the manufacture of high-profit speciality products which require high-quality iron. With the rise in iron prices during World War I production increased again, and the coming of the Shijiazhuang–Taiyuan railroad opened new markets for some (unknown and probably short) time.

An economic survey of traditional industries in selected counties of Shanxi carried out in 1950 showed that large parts of the province were still heavily dependent on iron production, and in consequence were very poor. In the county of Changzhi 長治縣, for example, of 179 villages, 53 were 'coal and iron villages'. Before the War, coal and iron production had accounted for about 60 per cent of the income of the population of these villages. In 1950 production had fallen to about 30 per cent of the pre-war level, but still there were 444 crucible-smelting furnaces (*fanglu* 方爐) in operation. About 20,000 people were directly engaged in coal and iron production, and over 50,000 were directly or indirectly dependent on the industry for their livelihood. Similar conditions were found in several other counties as well.[28]

The economics of the iron market forced the ironmasters of Shanxi to adopt a poorer technology, which gave an inferior product; in the long run this meant the ruin of the Shanxi iron industry. In the Great Leap Forward, if only a method of controlling sulphur content had been known, the crucible

---

26. An earlier remark by Williamson (1870, p. 296) suggests that limestone was considered a normal part of iron production in Shanxi. In Pingding, in 1866, 'Natives told us that lime, coal, coal-charcoal, and a clay they called *kal* … are all found in the immediate neighbourhood.'
27. Shockley (1904, p. 852); Read (1921, p. 454); Tegengren (1923–24, pp. 323–324).
28. Anon. (*1954*, pp. 10–11, 81–82).

smelting process would have been attractive for the purposes of the campaign, for it required low investment and was easy to learn; but in fact it was abandoned and traditional blast furnace technologies were introduced from elsewhere in China.[29]

---

29. Yang Kuan (*1960*, p. 99).

7

Large- and small-scale ironworks in Guangdong

On the southern coast, in the mountainous subtropical province of Guang-dong,[1] the traditional iron industry had the peculiarity that it was divided into two distinct sectors. Hundreds of small blast furnaces rather like those of the Dabieshan region produced iron in small quantities for consumption in their immediate vicinity, while a number of very large blast furnaces in the mountains produced large amounts of iron for a very wide market. Most of the iron produced by the large-scale sector was shipped on the great rivers of the province to one centre, the industrial city of Foshan 佛山 (Fatshan), about 20 km from Guangzhou (Canton), where it was fabricated into products which were traded throughout China and Southeast Asia and even further afield.

Our sources for the iron industry of Guangdong have a greater historical depth than for any other part of China, allowing us to see it with some clarity as far back as the late Ming period. On the other hand our knowledge of the actual technologies used by the two sectors of this industry is very limited; as usual, however, it is with the technologies that we must begin.

Small blast furnaces

Our *only* source for the blast furnaces of the small-scale sector is the eight watercolour paintings reproduced here as Figures 21–28. These are the first eight of twelve watercolours in an album that is now preserved in the Bibliothèque Nationale, Paris ('Fer', C.E. Oe 119 in-4°, 1–8). The watercolours were painted by a Chinese artist in Guangzhou in the middle of the nineteenth century and acquired by the *Mission Lagrené*, a French

---

1. On the physical and human geography of Guangdong see Ye Xian'en & Chen Chun-sheng (1990); Imbault Huart (1899); Anon. (*1917*); Xu Junming (*1956*); Wu Yuwen (*1985*). On its economic history see Faure (1989); Zhu Jieqin (*1985*); Anon. (*1987b*).

Figure 21. 'Exploring the soil' (*tan tu* 探土).

Figure 22. 'Testing the iron [ore]' (*yan tie* 驗鐵).

Figure 23. 'Extracting the iron [ore]' (*qu tie* 取鐵).

Figure 24. 'Separating the lead' (*fen qian* 分鉛). This picture clearly shows the calcining of iron ore; the idea that lead is somehow involved is curious.

Figure 25. 'Purifying the dirt' (*qing ni* 清坭). The 'dirt' is the calcined ore.

Figure 26. 'Removing the fire' (*chu huo* 出火). This probably shows the slag from the blast furnace being disposed of. The water-colour has been reproduced in colour by Hauser (1974, p. 21).

Figure 27. 'Tipping out the iron' (*dao tie* 倒鐵).

Figure 28. 'Making bars' (*zuo tiao* 作條).

diplomatic and commercial mission which spent about a year in Guangzhou in 1844–45.[2]

Before showing the blast furnace the album illustrates exploration for suitable iron ore (Figures 21–22), the digging of ore from a hillside (Figure 23), and the calcining of the ore (Figure 24). The blast furnace is shown in Figure 25: it appears to have a cast-iron base (no doubt lined with fireclay) and a ceramic shaft. The shaft was probably made by plastering fireclay on the inside of a basket plaited in the intended form, producing the appearance of wickerwork. Figure 26, labelled 'removing the fire', probably shows a worker getting rid of slag which has been tapped from the furnace.[3] The tapping of iron from the furnace is shown in Figure 27: like the furnaces of the Dabieshan region, the whole furnace is tilted in order to pour out the molten iron into a ladle. Finally Figure 28 shows the iron being poured into simple bar-moulds and then being cooled off in water.

It is important to recognise that we do not know how well the artist knew the technical processes that he depicted, or whether his labels reflect the actual terminology used by the workers. The windbox seen in Figures 25 and 27 seems too small in comparison with the much larger ones used in the Dabieshan blast furnaces, which were about the same size. We might also wonder whether the tapping of the furnace was really done in such a difficult and dangerous way as that shown in Figure 27: compare the methods used in the Dabieshan region, shown in Figures 2, 3, and 4.

At a guess, the operation of this blast furnace was probably rather like that of the Dabieshan blast furnace, operating continuously for a period from a few days to a week or two, producing several hundred kilograms of cast iron per day. This was a very useful technology for the small-scale village ironworks which were to be found throughout the province of Guangdong.

Large blast furnaces

Our principal source for the large-scale ironworks of the province is an interesting passage in the *Guangdong xinyu* 廣東新語, 'New discourses on Guangdong', a book of jottings written about 1680 by Qu Dajun 屈大均

---

2. This information was kindly supplied by Mme Madeleine Barbin, Conservateur de la Réserve, Bibliothèque Nationale. The history of the mission is described by Lavollée (1853, pp. 1–304), who also gives an extensive bibliography of books and articles by members of the mission (pp. 415–417). Brief notes on this album are given by Huard & Wong (1966, p. 219); Cordier (1909, p. 246); Courant (1902–10, p. 179). On the Guangzhou watercolours in general see Daurand (1845, pp. 56–64); Feuillet de Conches (1856, pp. 256–260); Feuillet de Conches (1862, pp. 145–148); Lavollée (1853, pp. 361–364); Downing (1838, vol. 2, pp. 90–117); Crossman (1991, pp. 173–202 *et passim*); Clunas (1984).

3. Or perhaps, as Graham Hollister-Short has suggested to me, the picture might show the granulation of molten iron by pouring it into water.

(1630–96).[4] Chapter 15 concerns 'Products' (*huo* 貨), and one section of that chapter concerns 'Iron'. It is translated in Box 4. This text is clearly composed of information from several unrelated texts, and caution will be advisable as we use it in an attempt to understand the technology and economics of the traditional iron industry of Guangdong.

The passage begins with a discussion of a kind of iron ore; its interpretation would require help from a geologist. In the middle of this is the warning that an ore deposit, no matter how rich, is useless if there is not sufficient wood in the vicinity[5] (the only fuel used was charcoal), and in general the passage shows a constant awareness of the economics of iron production.

Further on we find a description of the work force of an ironworks: it includes more than 200 furnace tenders, 300 miners, and 200 'water-carriers' and charcoal producers; transportation is provided by 200 oxen and 50 river vessels. The blast furnace produces 2–4 tonnes of pig iron per day, and this is shipped down-river to Foshan. These figures are suspicious, for if we take them seriously we are forced to suppose that the cost of iron was enormous. Production alone, without transportation costs, seemingly required between 0·5 and 1 worker-year per ton of pig iron. Even the labour-intensive technology used in the Dabieshan region required far less labour than this, and production at this level would not require 50 ships to transport the product to market. Obviously the passage refers to a large firm which operated numerous ironworks scattered over a large area.[6]

The actual mining of the ore might have been done in the same general way as in the small-scale industry (Figure 23). Calcining is not mentioned, but seems a good bet that, since it was necessary in the small-scale industry (Figure 24), it was also practised in the large-scale industry. In the description of blast furnace operation there are at least two aspects which we should expect to see mentioned but do not: the addition of flux in the charge, and the tapping of slag. Fluxes can sometimes be omitted in blast furnace operation,[7] but there is no way at all to avoid the production of slag.

---

4. On Qu Dajun, see especially Hummel (1944, pp. 201–202); also Ou Chu (*1986*). There are several variants of parts of this passage in eighteenth-century Chinese texts, and some modern historians cite these in preference to the *Guangdong xinyu*, thus taking them as quotations from Qu Dajun's sources rather than from his text. See e.g. *Yue zhong jian wen* (1988 ed.), ch. 21, pp. 244–6; *Yue dong wen jian lu*, ch. *xia* 下, p. 131; *Nan Yue biji* 南越筆記, ch. 5, pp. 3a–5a; Li Longqian (*1981*, pp. 361–362, 363, 367); Luo Yixing (*1985*, p. 82).

5. A warning also given by Agricola in his *De re metallica* of 1556. Agricola (1912, p. 31).

6. Two charcoal blast furnaces would not normally be placed too close together, as this would mean a doubled load on forest resources with little or no gain in efficiency.

7. See first section of Chapter 5 above.

Box 4. Translation of a passage on iron-smelting by Qu Dajun 屈大均 (1630–96) in *Guangdong xinyu* 廣東新語, 'New discourses on the province of Guangdong', (1700 ed., ch. 15, pp. 7b–10a; cf. 1974 ed., pp. 408–410).

There is no better iron than Guangdong iron. In the iron-producing mountains of Guangdong, wherever there is yellow water seeping out, one knows there is iron. Digging there, one will find a large body of iron ore in the form of an ox; this is the 'iron ox'. If one follows the path of the underground water and digs deeply, more iron will be obtained.[1]

However, of the mountains which produce iron, it is only on those which are forested that one can operate a furnace. If the mountain is bare, even if there is a great deal of iron it will be of no use. This is why 'iron mountains' are not easy to find.

Splitting a body of iron ore layer by layer, one finds that each [layer] has a tree-leaf pattern. This differs on the two sides. If the mountain has a certain type of tree, then that tree's leaf pattern will be found in its iron ore. Even if one digs as deep as several tens of *zhang* 丈 [several times 32 metres], the same phenomenon is found. When it is extremely cold in south China, the leaves do not [normally] fall from the trees; it is only on mountains which produce iron that the leaves fall, and these are absorbed by the essence of the iron. This is an example of the Way of 'metal conquering wood'.

The iron ore has a spirit, and to this the furnace-master must sacrifice devoutly before he dares to operate a furnace. The furnace has the shape of a vase [*ping* 瓶] with its mouth upward. The breadth at the mouth is about a *zhang* 丈 [3·2 m]. The base is 3 *zhang* 5 *chi* 尺 [11·2 m] thick [*sic*!], and the height is half of that [5·6 m].[2] The thickness of the body is slightly more than 2 *chi* [64 cm]. It is built of ashes, sand, salt, and vinegar.[3] It is bound about with thick cane and braced with wood of the *tielimu* 鐵力木 and the *zijingmu* 紫荊木.[4] It is also built against a mountainside for greater solidity.

At the back of the furnace is an opening, and outside the opening is an earthen wall. The wall has two 'doors' [blower fans], 5–6 *chi* high [1·6–1·9 m] and 4 *chi* broad [1·3 m]. Four persons operate these 'doors', 'closing' and 'opening' [pushing and pulling] alternately in order to produce the force of the blast.

The two openings [the blast-hole and the tap-hole] are lined with 'water stone' [*shuishi* 水石, perhaps diatomite]. 'Water stone' is produced at Dajiang Mountain 大絳山 in Dongan 東安 District [modern Yunfu County 雲浮縣, Guangdong]. Its substance is not hard, and not being hard it does not 'receive the fire'. Not receiving the fire, it can endure long without altering; hence the name 'water stone'.

Furnace operation begins in the autumn and ends in the spring. Because the weather is cold the iron contains a great deal of water; metal is the source of water, and water flourishes in the winter, so that molten iron is engendered by cold.

When the iron ore is charged [*xia* 下] [into the furnace] it is mixed with 'hard charcoal' [*jian tan* 堅炭].[5] Usually a mechanical device [*ji che* 機車] is used to cast down [the furnace charge] from the mountain into the furnace.

---

1. The importance of underground water in the search for ores is also mentioned by the sixteenth-century writers Biringuccio and Agricola (Biringuccio (1942, p. 15); Agricola (1912, p. 116)).

2. *Di hou san zhang wu chi, chong ban zhi* 底厚三丈五尺，崇半之.

3. Presumably it is built of clay with these ingredients added.

4. *Mesua ferrea* and *Madhuca subquincuncialis*, 'common mesua' and 'peanut madhuca'.

5. This does not refer to mineral coal. Charcoal varies greatly in its mechanical properties; in a tall blast furnace it was essential that the charcoal be hard and strong so that it could support the weight of a high column of furnace burden.

The flames [from the furnace] light up the sky, and its dirty black smoke [*qi* 氣] does not disperse for several tens of *li* 里 [1 *li* = ca. 0·6 km].

When the iron ore has 'melted' it flows out into a rectangular mould and solidifies into an iron slab. In order to obtain it, the furnace is 'shaken' [? *jiao* 攪, 'struck'?] with a wooden pole [removing the clay which blocks the taphole], so that the molten iron flows out and is cast into another slab.

In twelve hours [24 modern hours] one slab should be produced each hour, with a weight of about 10 *jun* 鈞 [180 kg]. If two slabs are produced per hour, this is called a 'doubled cycle' [*shuang gou* 雙鈞]; then the furnace is excessively vigorous [*wàng* 王 = 旺], and in danger of damage. One must anoint the furnace with the blood of a white dog; then no accident will occur.

Among the Five Metals, Iron is the one which corresponds to Water; it is called the Black Metal [*hei jin* 黑金], and is formed by the Essence of Great Yin [*tai yin zhi jing* 太陰之精]. Its spirit is a woman. According to tradition the wife of a certain Mr. Lin 林, when her husband was in arrears in his official iron quota, threw herself into the furnace in order to make it produce more iron. Today those who operate the furnaces always sacrifice to her, calling her 'Madame Gushing-Iron' [*Yong tie furen* 湧鐵夫人]. Her story is bizarre in the extreme.

An ironworks generally has 300 families living around it, [providing services as follows:] furnace tenders, more than 200; miners, more than 300; water-carriers[1] and charcoal-producers, more than 200; pack animals, 200 oxen; freighters, 50 vessels. The expense [i.e. capital investment] of an ironworks totals not less than 10,000 *jin* 金.[2] If it produces more than 20 slabs per day it is profitable, while eight or nine slabs [per day] is unprofitable; this is a fundamental principle.

Of all the smelters it is the iron of Datangji Furnace in Luoding 羅定大塘基爐 which is best. All of it is 'first grade iron' [*kai tie* 鎧鐵], glossy and soft, which can be drawn into wire. When it is cast into woks [*huo* 鑊][3] it is also good and 'hard' [*jian* 堅]. It is more expensive than the first-quality [iron] of any of the other smelters.

After smelting at any of the smelters, all of the iron is shipped to the port city of Foshan 佛山, where the people are excellent founders. Of the woks which they produce, the largest are called '*tangwei* 糖圍' ['sugar pot'?], 'deep-seven', 'deep-six', 'ox-one', and 'ox-two'. The smallest are called 'ox-three', 'ox-four', and 'ox-five'. If there are five to a bundle they are called 'five-*kou* 口', and if there are three to a bundle they are called 'three-*kou*'. Those without handles are called 'oxen'. The best are called 'pure'.

In former times those who cast [woks] without handles were not permitted to cast them with handles, and those who cast them without handles were not permitted to cast them with handles. Those who cast both were indicted.

After [a wok] is cast it is coated with yellow clay and pork fat [to prevent rust].

When [a wok] is struck with a light stick, if it [sounds] like wood then it is good. It sounds like wood because its substance is 'hard' [*jian* 堅]. Thus the woks of Foshan are expensive, because they are 'hard', and those of Shiwan 石灣 [in modern Boluo

---

1. *Jizhe* 汲者, perhaps to be understood more generally as 'fetchers and carriers'.

2. In this type of context one *jin* usually means one *liang* 兩 (ca. 37 grammes) of silver, so that the sum indicated, if it is to be taken literally, amounts to 370 kg of silver.

3. The Cantonese pronunciation of this word gives us the English word *wok* for the round-bottomed pan which is a fundamental implement in every Chinese kitchen. The more usual north Chinese word is *guo* 鍋.

County 博羅縣, Guangdong] are cheap, because they are brittle.[1] When they are sold [as far away as] central China, people can distinguish them by their thinness and smoothness; the foundrywork is refined, and the craftsmanship is accomplished. Foundry products are generally better in Foshan, while ceramic [products are better] in Shiwan.

In iron-fining [*chaotie* 炒鐵], cast iron is kneaded [*tuan* 團] and charged into the furnace [i.e. the fining hearth]. It is fired until it glows red,[2] then taken out and put on the anvil. One person holds it with tongs and two or three hammer it. At the side ten or more youths [*tongzi* 童子] fan it [i.e. work the bellows for the blast in the fining hearth]. The youths always sing a chanty without stopping. After this [the iron] can be converted into wrought iron and made into thin plates.

[In Foshan?] there are several tens of iron-fining plants [employing] several thousand persons. Each plant has several tens of anvils and at each anvil there are more than ten persons.

This [the fining hearth] is the 'small furnace'. Furnaces are large or small according to whether the iron [produced] is cast or wrought. The treatment [*zhi* 治] of cast iron is a matter for a 'large furnace' [a blast furnace], while the treatment [zhi 治][3] of wrought iron [i.e. the conversion of cast iron to wrought iron] is a matter for a small furnace.

The excellence of the strength [*jian* 健][4] of steel lies in its quench-hardening [*cui* 淬]. Before it has been quench-hardened, its softness remains. In quench-hardening, the steel is first hammered [into the desired shape] at the forge, then removed from the fire and put into water. A great fire is needed to soften it, and pure water is needed to strengthen it and form pure steel. This is the refining of steel [*lian gang* 煉鋼].

Ganquan[5] 甘泉 has said: Observe the casting of metal in a great furnace, and you will know the end and beginning of Heaven and Earth. When [the metal] is melted in the furnace, this is its birth. When it leaves the furnace and solidifies, this is its accomplishment. Melting is a matter of the watery beginning, and solidification is a matter of the earthy end. Men consider melting to be subjugation, but do not realise that it is the beginning of birth; what could be more enduring? Men consider solidification to be enduring, but do not realise that it is the end of accomplishment; what could be more of a subjugation? 'Beginnings and ends alternate, subjugation and endurance interact',[6] but the metal never changes; this is an image of the Way.

---

1. The point of this passage is that the best woks will be of grey rather than white cast iron. White cast iron is very hard and can ring like a bell, while grey cast iron is soft and contains microscopic flakes of graphite which deaden vibration. Both are brittle (for very different reasons), but white cast iron is considerably more brittle than grey. The term *jian* 堅 normally means 'hard', and is so translated here, but it might be better to take it as meaning 'strong' or 'robust' in a vague sense. In modern technical terms, the mechanical property which is relevant here is *toughness*.

2. We should expect the iron to be white-hot rather than red-hot when taken from the furnace.

3. Perhaps the two occurrences of *zhi* 治, 'to order, govern' in this sentence are scribal errors for *ye* 冶, 'to smelt', though in the second case this would be an incorrect use of the word.

4. Here the relevant mechanical property is in fact *hardness*.

5. Possibly the philosopher and educator Zhan Ruoshui 湛若水 (1466–1560), whose literary name was Ganquan. He came from Zengcheng 增城 in Guangdong, and wrote a number of books which, judging from their titles, might well contain the passage quoted by Qu Dajun here. See Goodrich & Fang (1976, pp. 36–42). Or conceivably the person referred to might be Qu Dajun's son Qu Minghong 屈明洪, whose alternate name was Ganquan; see Hummel (1944, p. 202).

6. *Qu shen xiang gan* 屈信相感, a quotation from the *Xici* 繫辭 commentary of the *Book of Changes*. *Shisan jing zhushu*, ch. 8, p. 87c; Legge (1882, p. 389).

The text mentions that furnace operation begins in the autumn and ends in the spring, and explains that this is because the element Water is in the ascendant in the winter, so that melting is facilitated. In eighteenth-century Britain, before the invention of the hot blast in 1828, it was a widely-remarked phenomenon that blast furnaces operated more efficiently in winter than in summer. The reason for the greater efficiency in winter was in fact that the air contained *less* water.[8]

The description of the blast furnace itself mentions a base, 11·2 m thick. We know from Han archaeology, and from accounts from the Great Leap Forward,[9] that large blast furnaces required a substantial base which could withstand the high temperatures at the bottom of the furnace and prevent any moisture from penetrating into it. It was typically made by digging a deep hole in the ground which was filled up with alternating layers of loose stones and tamped clay. A thickness of 11·2 metres for this base is surely excessive, but there is no obvious alternative interpretation for this point. We should not try to make too much of it, but assume for the moment that it reflects some misunderstanding, either mine or the author's.

The height of the blast furnace is said to be half of the given thickness of the base, or 5·6 metres. This is more in line with what we otherwise know about large Chinese blast furnaces, though we might have expected it to be even higher. The blowing apparatus is not the more common 'wind-box', or double-acting piston bellows, but a type with two large hinged fans which are pushed and pulled alternately by four workers. It is a surprise that the blast is labour-powered rather than water-powered, but it may be that geographical considerations made this difficult in the ironworks about which Qu Dajun had information.

The mention of a mechanical conveyance for the furnace charge is interesting. Probably this device carried the charge up to an appropriate height, then was arranged to 'cast it down' into the furnace mouth from a sufficient distance to the side so that the apparatus, presumably made of wood, was safe from the flames shooting out of the furnace.

Qu Dajun states that the best iron in Guangdong comes from a place called Datangji 大塘基 in Luoding Department 羅定州. In 1978 and 1982 investigations were carried out here by the Guangdong Provincial Museum, and some useful information came to light.[10]

The ruin of a blast furnace believed to be of the early Qing period was investigated in 1978 in a village named Luxia 爐下, 'Below the Furnace'. It is built into the side of a mountain near a stream, and the ruin of a 'water-powered pounder' (*shuidui* 水碓), believed to have been used in ore dress-

---

8. Percy (1864, p. 397).
9. See e.g. Yang Kuan (*1960*, pp. 132, 136).
10. Cao Tengfei & Li Caiyao (*1985*); Cao Tengfei & Tan Dihua (*1985*, pp. 118–123).

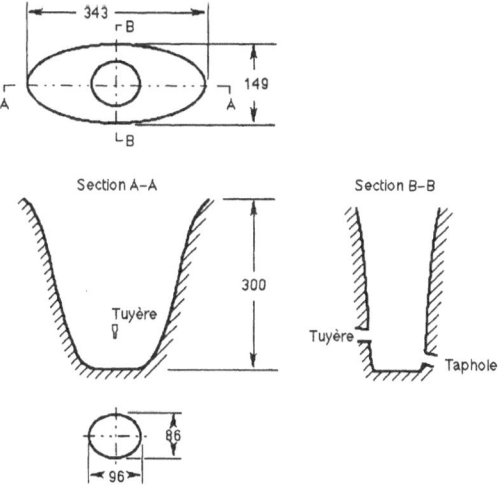

Figure 29. One possible reconstruction of an early Qing blast furnace investigated at Lu-xia Village in Luoding County, Guangdong, redrawn from Cao Tengfei & Li Caiyao (*1985*, p. 72, fig. 5). Dimensions are given in centimetres. The figure does not corre-spond well to the description in the text of the article, according to which the entire north wall of the furnace above the taphole is missing, the remaining height of the south wall is 271 cm, and the original height was 6–8 m. The measurements shown here for the mouth of the furnace are those of the remaining upper part of the furnace as excavated.

ing, lies a half kilometre upstream. Numerous old mine-pits, called *geng-longtou* 埂䃏頭 by local people, are found on the mountain, and there is still virgin forest not too far away. Large amounts of slag and charcoal turn up in the fields around the furnace site. The Luxia furnace has enough points in common with the description in *Guangdong xinyu*, including its time and place, that it is quite likely that Qu Dajun (or the author of his source) had actually seen a furnace like this one.

The furnace in Luxia is built on a base of refractory bricks laid on a 'very thick' layer of clay mixed with salt and sand. The height of its charging platform indicates that the furnace was 6–8 m tall; the height of the ruin is 2·71 m. The investigators give the diagram shown in Figure 29, which does not entirely match their verbal description. The walls of the furnace are 77 cm thick. A great surprise here is that the furnace appears to have had an elliptical cross-section. Some elliptical blast furnaces are known from Han China, and this form was also tried in nineteenth-century America, Britain, and Russia,[11] but the purpose in these cases was to distribute evenly the blast from four or more tuyères, which were placed on the long walls of the furnace. In this case, with only one tuyère, it is not clear what purpose the elliptical shape served.

11. Percy (1864, pp. 489–491).

There are numerous blast-furnace ruins in modern Luoding County, usually in villages with the character *lu* 爐, 'furnace', in their names. There are two places with the name Datangji (mentioned by Qu Dajun), but there is no sign of iron production in the immediate vicinity of either, and we must suppose that one or both of these served as trans-shipment points for iron from the scattered ironworks of the area, which perhaps all belonged to the same firm. Both are on the Luojing River 羅鏡河. This flows into the Shuangshui 瀧水, which flows into the Xijiang 西江 (West River), one of the major rivers of Guangdong, which flows into the sea near Guangzhou.

> ... In the past, commercial vessels could navigate the Luojing River as far as the wharves at Fenjie 分界; today it is no longer navigable because of stones and shallow water. ... [One of the places named Datangji is in Fenjie.] Here there is a great bay which is suitable as a harbour for commercial ships. According to several retired sailors, including Yang Ya 楊呀, Xie Wenda 謝文大, Chen Shengcai 陳生才, and Zhou Jin 周金, in the past there were normally 30 to 40 ships berthed in this bay. These ships carried pig iron, iron woks, mountain products, and other local products down the Luojing River, the Shuangshui, and the Xijiang as far as Zhaoqing 肇慶, whence the loads were carried further to Foshan, Guangzhou, and other places. On the return trip they carried handicraft products from other areas as well as scrap iron ... The older generation of sailors recalled the hard-working life of sailing on the Luojing River so vividly that we could see it before our eyes. Boatmen who served their apprenticeship under Yang Ya and the others are still working for the Luoding Wooden Sailing Ship Transport Company 羅定木帆船運輸社.[12]

It is something of a surprise to learn that iron was being produced somewhere in this vicinity within living memory, for we seem to have no written sources indicating that iron was produced here in the nineteenth or twentieth century.

Six ruined furnaces were investigated in 1982, in villages named Tielu 鐵爐, Jiuludu 舊爐督, Jigonglu 雞公爐, Lechalu 竻揸爐, Zaoshilu 鑿石爐, and Shuiyuanlu 水源爐. According to a retired schoolteacher in Tielu, Ye Qihua 葉其華, these were all owned by one man, named Mai Wenyuan 麥文元. Iron from all six furnaces was transported to Tielu, where conversion to wrought iron took place. A fining or puddling hearth (*chaolu* 炒爐) was indeed found near the blast furnace at Tielu. We are not told what sources Ye Qihua had his story from, but no doubt he had access to local records, family documents, and the like. Perhaps also he had interviewed descendants of Mai Wenyuan. It is a pity that no dates are given for this industrialist, but presumably he lived in the early Qing period, when many sources indicate that the iron industry flourished here.

◆ ◆ ◆

---

12. Cao Tengfei & Li Caiyao (*1985*, pp. 70–71).

Other ruins of early Qing blast furnaces were noted by Yang Dajin in 1938 at Wushiling in Yunfu County 雲浮烏石嶺, some 65 km east of Luoding:

> The ore here was discovered very early: in the Qianlong 乾隆 period [1736–95] it was already being mined by the local people, and the remains of their diggings are numerous. Their products were iron woks, old-style cannons, large temple bells, large incense burners, etc. The iron-casting furnaces are still there; for example the ruins of Daping Furnace 大平爐 and Daan Furnace 大安爐. In form they somewhat resemble modern lime kilns. The fuel used in their operation was taken from nearby mountain forests. When this fuel was used up, furnace operation was stopped, for the difficulty of transportation made it extremely uneconomical to bring in fuel from elsewhere. Another reason for discontinuing mining here was geomantic superstition.[13]

The description of these furnaces as resembling modern lime kilns makes it likely that they were of the same form as the Luoding furnaces. It is generally believed that all or nearly all of the iron from these *dalu* was shipped to Foshan for further processing, but the range of products listed here indicates that the situation must have been more complex than this.

The iron industry of Guangdong

The debate of recent decades on 'embryonic capitalism'[14] in late Imperial China has included a large amount of research, some of it first-rate research on topics in Chinese economic history which for too long have been ignored. An amazing mass of source material has been uncovered, and numerous high-quality articles published.[15] Much of this source material is not easily available outside Guangdong, and the following discussion will be more dependent than usual on quotations in secondary sources.

Qu Dajun mentions two types of furnace, large and small, *dalu* 大爐 and *xiaolu* 小爐, and from his explanation it is clear that these are respectively blast furnaces and fining or puddling hearths (Box 4). A third type of furnace often mentioned in the sources is the *tulu* 土爐, which presumably is to be taken as a 'local furnace', a furnace to serve local needs. The *tulu* were smaller than the *dalu* and were used in two ways: as blast furnaces, to smelt iron ore,

---

13. Yang Dajin (*1938*, p. 352).
14. *Zibenzhuyi mengya* 資本主義萌芽, sometimes translated literally as 'sprouts of capitalism'.
15. Important publications on the iron industry of Guangdong include Anon. (*1983*, pp. 493–496); Li Longqian (*1981*); Luo Hongxing (*1983*); Luo Yixing (*1984*); Peng Jianxin (*1994*); Deng Kaisong (*1985*); Tan Dihua (*1987*); Cao Tengfei & Tan Dihua (*1985*); Wang Hongjun & Liu Ruzhong (*1980*); Ye Xian'en (*1987*). In Western languages, see Hirth (1890); Eberstein (1974, pp. 23–60 *et passim*).

and as cupola furnaces, to melt pig or scrap iron for casting.[16] It is likely that the small blast furnace shown in Figures 25 and 28 is precisely such a *tulu*.

Both the *dalu* and the *tulu* were blast furnaces which produced cast iron from ore, but they were supervised by different departments of the Qing government. The regulations are quoted in the premises of the judgement in a law case of 1821 on the illegal opening of an ironworks:

> There is a distinction between *dalu* and *tulu*. The *dalu* cast[17] iron ingots. They are given licences by the Provincial Treasurer, who reports to the Provincial Governor and he to the Ministry. The furnace levy (*luxiang* 爐餉) is collected by and goes into the account of the Provincial Treasury. The iron tax (*tieshui* 鐵稅) is collected by the Salt Controller and also transferred to the Provincial Treasury. The *tulu* cast[18] farm tools, and are licensed by the Salt Controller, who reports to the Governor-General and he to the Ministry. Both the furnace levy and the iron tax are collected by the Salt Controller and transferred to the Provincial Treasurer.[19]

Presumably the Provincial Treasury had the technical expertise necessary to oversee the large-scale ironworks with *dalu*, while the Salt Controller had a finer net of agents and was therefore better able to deal with the much more numerous, but administratively less important, *tulu*. The 'furnace levy' probably had its origin in a tax in kind, but we see it in the Qing sources as an annual tax in silver paid without relation to actual production. We know very little about the 'iron tax',[20] but quite a bit about the furnace levies. Li Longqian has compiled from diverse sources a table of ironworks established in Guangdong in the Qing; his necessarily incomplete table lists 87 *dalu* and

---

16. Deng Kaisong (*1985*, pp. 183–184); Li Longqian (*1981*, pp. 355, 364–366). Cao Tengfei & Tan Dihua (*1985*, p. 118) are mistaken when they state that the *tulu* was used for converting cast iron to wrought iron: they cite as evidence for the statement the passage translated immediately below from *Yue dong cheng'an chubian*, but in fact it states clearly that the *tulu* was used for casting agricultural implements.

17. *Zhushan* 鑄煽, probably 'to cast from a blast furnace'.

18. *Zhuzao* 鑄造, simply 'to cast'.

19. 有大爐、土爐之分。大爐鑄煽鐵斤，由藩司給照詳明巡撫衙門咨部，爐餉由藩司征收報撥，鐵稅運司征收轉解藩司報撥。土爐鑄造農具，由運司給照詳明總督衙門咨部，餉稅均由運司征收轉解藩司。*Yue zhong cheng'an chubian*, ch. 24, p. 28b; also quoted, with a typographical error, by Cao Tengfei & Tan Dihua (*1985*, p. 118). The legal jargon of the passage is difficult, and I have been glad to have help from my friend Charles Curwen. The law case is discussed in some detail by Li Longqian (*1981*, pp. 364–366).

20. According to Cao Tengfei and Tan Dihua (*1985*, pp. 127–128) the tax was in the early Qing 0·8 *liang* of silver per 10,000 *jin* of cast iron and 1·2 *liang* per 10,000 *jin* of wrought iron. Later this was changed to 1 *liang* per 10,000 *jin* of any kind of iron (37 grams silver per 6 tonnes iron).

33 *tulu*.[21] The actual amounts of the furnace levies are available for 47 *dalu* and two *tulu*. Of these the vast majority of the *dalu* levies were 53 *liang* 兩 of silver, nearly all of the rest paying 50. The two *tulu* levies were both only 5·3 *liang*.

Thus there were two distinct technologies for primary iron production in use in Guangdong in recent centuries, and the ironworks using them were recognised in law as forming two distinct sectors of the iron industry. Ironworks of the small-scale sector produced pig iron in small blast furnaces, and used the same furnaces to melt pig iron or scrap for casting products for local consumption. The large-scale sector was made up of large firms running numerous ironworks in the forested mountains of the province. In these ironworks large blast furnaces produced pig iron. Most of their production was shipped by river to the industrial town of Foshan, near Guangzhou, though as we have seen, some was cast or converted to wrought iron at the works for local consumption.

In Foshan some of the pig iron was converted to wrought iron, while the rest was used by foundries.[22] In either case the products were marketed far and wide: up and down the coast of China and all over Southeast Asia.[23]

In 1890 Friedrich Hirth called Foshan 'the city of iron and steel wares', and estimated its population as close to a million, 'mostly of the working class'.[24] We have no Western travellers' accounts of iron production here, for these industrial workers were considerably less tolerant of foreigners than the more commercially oriented people of Guangzhou. In 1847 a group of Englishmen and Americans visited Foshan and were attacked by an angry crowd which they claimed numbered twenty to thirty thousand. They were rescued by a company of Chinese soldiers ordered out by the city magistrate.[25]

---

21. Li Longqian (*1985*, pp. 372–379).

22. On Foshan and its iron industry see especially Luo Hongxing (*1983*); Luo Yixing (*1984*; *1985*); Tan Dihua (*1987*); Anon. (*1987a*); Wang Hongjun & Liu Ruzhong (*1980*); Faure (1990); Xian Jianmin (*1993*); Huang Jianxin & Luo Yixing (*1987*); Jiang Zuyuan (*1987*); Imbault Huart (1899, pp. 25–27).

23. On the trade between Guangzhou and Southeast Asia, see especially Bronson (1992); Viraphol (1977); Yu Siwei (*1983*). Bronson (1992, p. 94) gives statistics from Dutch colonial records for metal imports in Java in the years 1679–81, and these seem to show that the early Qing prohibition of the export of iron applied to wrought iron but not cast iron.

24. Hirth (1890, p. 96).

25. *Chinese repository*, March 1847, **16**: 142–147. A rumour in the local English-language press had it that one Westerner had spent a day and a night in Foshan, disguised as Chinese (*Chinese repository*, Feb. 1846, **15**: 64–65, quoting *Hongkong register*). This may possibly have been the Swedish commercial attaché C. F. Liljevalch, who seems slightly better informed than other writers on the Guangdong iron industry, but even his report is too brief to be useful here (Liljevalch (1848)).

A memorial by E Mida 鄂彌達 in 1734 stated that there were at least 50 or 60 ironworks in mountainous regions of Guangdong, employing several tens of thousands of workers.[26] Another source, a century later, states that the annual production per furnace is 800,000–900,000 *jin* 斤.[27] Some modern historians multiply 60 furnaces by 900,000 *jin* and on this basis reach an estimate of annual production for the province of 54,000,000 *jin*, or about 32,000 tonnes.[28] Clearly this method necessarily overestimates total production, and we should be justified in reducing it by at least half. In 1735 E Mida stated in another memorial that there were iron-smelting furnaces in every department and district (*zhou* 州 and *xian* 縣) of the province;[29] taken literally, this would imply well over 100 furnaces, and not all in mountainous regions. Obviously the 1734 memorial referred to *dalu* and the 1735 memorial to *tulu*.

We have seen that the iron industry in Guangdong was largely overseen by the same government departments which dealt with the production and distribution of salt. A treatise of 1762 on the salt administration of Guangdong, *Liang Guang yanfa zhi* 兩廣鹽法志, includes a chapter on the iron industry which gives some interesting information. After a brief history going back to the pre-Qin period, it gives some statistics on production of iron in the Qing. The text is difficult, but it can be interpreted as giving total taxed production per year:

> From the first establishment of the Dynasty to the present there have been [years in which the tax on] 7,139,000 *jin* 斤, 6,839,000 *jin*, and 5,892,000 *jin* was collected [*zheng* 徵]; though the amount varies it is normally possible to obtain [the tax on] more than 6,000,000 *jin*. This is the normal amount. Thus the iron tax of Guangdong can be said to be plentiful.[30]

Six million *jin* is about 3,600 tonnes. This appears to be distinctly on the low side, and may indicate that a good deal of the tax was evaded.[31] At the end

---

26. The memorial is quoted *in extenso* in Anon. (*1983*, pp. 32–33). Also quoted by Li Longqian (*1981*, p. 362) and Luo Yixing (*1985*, p. 84).

27. The 1835 edition of *Liang Guang yan fa zhi* 兩廣鹽法志 (which I have not seen), quoted by Luo Yixing (*1985*, p. 84).

28. E.g. Luo Yixing (*1985*, pp. 84, 87).

29. Memorial by E Mida 鄂彌達, quoted by Li Longqian (*1981*, p. 362).

30. 我朝自定鼎以迄於今有徵　[7,139,000]　斤者有徵　[6,839,000]　斤者有徵 [5,892,000]　斤者雖參差不等大抵可得六萬餘斤此其恆也是廣東一省之鐵稅可謂盛矣。
    Liang Guang yan fa zhi (1762 ed.), ch. 24, p. 2a.

31. The figure is low if all of the 45 blast furnaces mentioned in the following were *dalu* producing several hundred tonnes per year each. The amounts stated cannot be the actual tax collected, since the tax was not in kind but in silver, and such tonnages of silver would be utterly unthinkable.

of the treatise chapter is a list of blast furnaces in Guangdong. Of these 45 were in operation, 19 were temporarily out of operation, one was out of operation because of bankruptcy, and eight had been abandoned.

The decline of the Guangdong iron industry which began in the late eighteenth century has been documented by several Chinese historians.[32] One of many indications is that the 1835 edition of the *Liang Guang yanfa zhi* lists only 25 *dalu* in operation in 1799,[33] compared with the 45 listed in the 1762 edition. A memorial by the famous reformer Zhang Zhidong 張之洞 (1837–1909) in 1889 describes the traditional Guangdong iron industry as being in deep decline, and recommends various policy changes to make it more competitive with foreign imports.[34] By this time the large *dalu* were all gone, but the small *tulu* continued to be used. In 1954 they continued in use in at least three counties of Guangdong, and had recently been reintroduced in several others.[35] Many more were built all over the province in connection with the Great Leap Forward of 1958.[36]

---

32. E.g. Luo Yixing (*1985*, pp. 90ff).

33. Quoted by Luo Yixing (*1985*, p. 90); cf. Li Longqian (*1981*, pp. 372–378).

34. Zhang Zhidong (*1937*, ch. 27, pp. 1a–4a). The whole memorial is interesting and would repay closer attention. On Zhang Zhidong see Hummel (1944, pp. 27–32); Bays (1970).

35. They continued in use in Zijin 紫金, Pingyuan 平遠, and Xinyi 信宜, and were newly reintroduced in Liannan 連南, Lianjiang 廉江, and Huaxian 化縣. Xu Junming (*1956*, p. 104).

36. Peng Jianxin (*1994*).

8

Concluding remarks

In the period before the massive influence of Western commerce the different Chinese iron industries of this discussion appear to have fit very well in their particular geographic contexts. In the Dabieshan region, where transportation difficulties made it uneconomic to carry iron any distance, small blast furnaces provided for the needs of consumers in their immediate vicinity. In the Sichuan basin, which was relatively isolated from the rest of China but had very good intraregional transportation, iron was smelted in large blast furnaces, which gave significant economies of scale. The rugged mountains of Guangdong meant that in numerous areas it was economic to produce iron for local markets in small blast furnaces, but at the same time the great rivers of Guangdong, and the opportunity to export products by sea, meant that the province could also support a large-scale iron industry producing iron in large blast furnaces.

The iron industry of Shanxi produced for a very large market in north China, but its firms were all very small. This is presumably a consequence of the particular technology which was used here, crucible smelting, which seems not to provide significant economies of scale – or perhaps it is a mistake to believe that von Richthofen's description of the Shanxi iron industry in 1870 tells us much about the situation before it began the decline which he so graphically describes. Was the technology the same a hundred years before? Was the typical size of firms the same? We may yet be lucky enough to find written sources relevant to the second question, and there is surely archaeological material waiting in the ground to help with the first.

The few cases in which it has been possible to catch glimpses of the workers, businessmen, and bureaucrats involved should serve to remind us that technologies and industries are made by people, not by impersonal forces of Progress or Economic Equilibrium. Nevertheless, to the extent that we can see the industries before modern times at all, they appear to be in dynamic equilibrium with their surroundings. Each uses an 'appropriate technology' – appropriate to its endowment of raw materials, to its labour force, and to its market and the associated transportation possibilities.

◆ ◆ ◆

Western contact brought, in the nineteenth century and even earlier, significant disturbance to this equilibrium. Very early on it seems that new investment opportunities in foreign trade led to a decline in investment in the large-scale sector of the Guangdong iron industry. From the early nineteenth century Western iron could compete successfully in Chinese markets, and this further weakened the large-scale iron industries of China. The social and political consequences of this decline need research: remember for example that many of the early leaders of the Taiping Rebellion (1851–64) were unemployed miners from Guangdong and Guangxi.

The large-scale industry was sensitive to disturbance by Western contact because large-scale production requires large markets, large markets require good transportation, and regions with good transportation are normally the first to be penetrated by outside competition. Sichuan was an exception, a large market which could not be penetrated by imports of a relatively cheap commodity until steamship traffic on the Yangzi had matured in the 1920s.

The small-scale industry survived much better, and even expanded in some places, taking over some of the peripheral markets of the large-scale industry. This industry was more robust because it required less capital and could rely on a labour force with few other opportunities for employment. We do not see in the sources any sign that the small-scale technologies of Dabieshan and Guangdong changed in significant ways.

In Sichuan, on the other hand, the large-scale iron industry responded to new economic conditions by adapting its technology to make it less capital-intensive and more labour-intensive. In one sense these developments represented a technological degeneration; in another, more detached, sense they may be seen as progressive, for they made for a more efficient iron industry in the specific context in which it functioned. In dealing with technological change it is always well to remind oneself of these two sides of the question of progress.

It is not clear that the technological changes in Sichuan had any very serious effect on the quality of the iron produced. In Shanxi, on the other hand, technical changes made in response to the new economic conditions degraded the product: by the beginning of the twentieth century Shanxi iron contained so much sulphur that it was decidedly inferior to what could be obtained elsewhere. Shanxi could compete only by offering extremely low prices, thus exacerbating the problems and laying the basis for the province's continuing poverty today.

War and revolution in the twentieth century brought continuous change. The sudden loss of imported iron in World War I brought soaring prices and new opportunities for the traditional iron industry. The small-scale industries of places like Dabieshan flourished, producing for large markets in which, almost unbelievably, iron was transported hundreds of kilometres on the backs of coolies. In such a dire economic situation the large-scale industries might have had an even better chance to expand and prosper, but the technology by which they had flourished earlier was by this time largely forgotten.

From the end of the nineteenth century onward was for China a time of intense industrial development. This usually meant the jettisoning of traditional technologies and their replacement with Western technologies, but in the iron industry there were some exceptions to this rule. In the 1930s the warlord Liu Xiang in Sichuan appears to have encouraged the improvement of the traditional iron-production technology through the application of modern scientific advice; whether this actually helped must be a matter for future research. During the Anti-Japanese War, the Guomindang government in Sichuan and the Communist government in northern Shaanxi were both largely dependent on traditional iron-production technologies, and worked to improve them. A question for future research is what the Japanese occupation government did about the traditional iron industry in the parts of China which it controlled.

After the Revolution of 1949, with the coming of peace, political stability, and national reconstruction after decades of war, the demand for iron and steel increased enormously, and China's modern sector could not expand rapidly enough to satisfy it. In many places in the early 1950s, especially in the lower Yangzi valley, small traditional ironworks were established by local authorities to meet local requirements. Compared with modern ironworks these had both advantages and disadvantages. They required a much smaller initial investment and were built using local materials. The necessary skilled workers were already available, and their skills could be acquired on the job fairly quickly by new workers. Advantage could be taken of small sources of raw materials which could not be exploited economically by modern large-scale methods. The problems of social dislocation associated with large-scale industry could be avoided. The burden on China's still-inadequate transportation facilities could be alleviated. And it must have been a great advantage to have a local supply of iron, for reliance on outside supplies must have involved difficult political negotiations.

There were serious disadvantages to the use of these small ironworks, in the cost of the product and often also in its quality; but on balance it seems that the temporary expansion of the traditional sector was a sensible solution to some pressing problems which confronted China in the 1950s. But then came the Great Leap Forward of 1958–60 and the related campaign to build hundreds of thousands of small-scale ironworks all over China. It was a great fiasco: enthusiasm for an idea was in command, and the voices of realists, though audible, were not heard. Nevertheless it is important to notice the partial success which the campaign had in 'backward' isolated regions, where iron production remained a living tradition and where high unemployment meant that the opportunity cost of labour was very low. Had political conditions been different, an expansion of iron production in these regions might have played as important a role in China's economy in the 1960s as it had done during World War I. Much research remains to be done on the campaign for iron and steel production of the Great Leap Forward, but if this research is to go beyond political posturing it must be done with a background in the actual technologies used.

Bibliography

Chinese works from before 1800 are listed by title, all others by author and year. Works by corporative authors are listed under 'Anon.' The year is italicised if the work is in Chinese or Japanese. Most translations of Chinese and Japanese titles are my own; those translations which are taken from the publication itself are placed in inverted commas.

Abendanon, E. C. (1906). 'La géologie du Bassin Rouge de la province du Se-Tchouan (Chine).' *Revue universelle des mines, de la métallurgie, des travaux publics, des sciences et des arts appliqués à l'industrie: Annuaire de l'Association des Ingénieurs sortis de l'École de Liège*, 1906, **14**, 225–243; **15**, 34–125, 237–287; **16**, 61–98.

Adshead, S. A. M. (1984). *Province and politics in late Imperial China: Viceregal government in Szechwan, 1898–1911*. Curzon Press, London and Malmö, 1984. (*Scandinavian Institute of Asian Studies monograph series*, 50).

Agricola, Georgius [Georg Bauer] (1912). *De re metallica: Translated from the first Latin edition of 1556 with biographical introduction, annotations and appendices upon the development of mining methods, metallurgical processes, geology, mineralogy & mining law from the earliest times to the 16th century*. Translated by Herbert Clark Hoover and Lou Henry Hoover. The Mining Magazine, London, 1912. Facs. repr. Dover Publications, New York, 1950.

Alley, Rewi (1961a). *China's hinterland—in the leap forward*. New World Press, Peking, 1961.

Alley, Rewi (1961b). *Together they learnt how to make iron and steel. Some early types of furnaces used in 1958–9, in China*. An unpublished album of 299 photographs; this and most of the original negatives are in the collection of the Needham Research Institute, Cambridge.

Anon. (1834). 'Imports and exports of Canton.' *Chinese repository*, 1834, **2** (no. 10), 447–472. 'The principal authors consulted were Crawfurd's Indian Archipelago, Milburne's Oriental Commerce, Macculloch's Commercial Dictionary, and Hooper's Medical Dictionary. Considerable aid was also obtained from merchants in Canton.'

Anon. (1910). 'Native coal and iron working in the province of Shansi, China.' *Engineering*, 1910, 761–762 + plates 81–82.

Anon. (*1917*).
　　Shina shōbetsu zenshi. 1: Kanton shō. Fu Honkon, Makao 支那省別全誌
　　第一卷　廣東省　附香港澳門.
　　Complete gazetteer of China. Vol. 1: Guangdong, with Hong Kong and Macao.
　　Tō-A Dōbunkai 東亞同文, Tōkyō, 1917.

Anon. (*1920*).
　　Shina shōbetsu zenshi. 17: Sansei shō 支那省別全誌　第十七卷　山西省.
　　Complete gazetteer of China. Vol. 17: Shanxi.
　　Tō-A Dōbunkai 東亞同文會, Tōkyō, 1920.

Anon. (*1922*).
　　Shina shōbetsu zenshi. 5: Shisen shō 支那省別全誌　第五卷　四川省.
　　Complete gazetteer of China. Vol. 5: Sichuan.
　　Tō-A Dōbunkai 東亞同文會, Tōkyō, 1922.

Anon. (1939). *Metals handbook.* American Society for Metals, Cleveland, Ohio, 1939.

Anon. (1944). *China proper. Vol. 1: Physical geography, history and peoples.* Naval Intelligence Division, 1944. (*Geographical handbook series*, B. R. 530). Largely written and edited by P. M. Roxby.

Anon. (*1954*).
　　Shougongye ziliao huibian 手工業資料彙編 1950–1953
　　Collected materials on handicraft industry, 1950–1953.
　　Comp. by the Institute of Economics, Chinese Academy of Sciences 中國科學院經濟研究所.
　　Chinese Academy of Sciences, Beijing, 1954 (*Jingji wenti cankao ziliao* 經濟問題參考資料, 1).

Anon. (*1958a*).
　　Tufa liantie 土法煉鐵.
　　Regional [Henan] methods of iron production.
　　Henan Office of Metallurgy, Zhengzhou, 1958.
　　Tr. Wagner (1985, pp. 5–27).

Anon. (*1958b*).
　　Tufa diwen lian'gang 土法低溫錬鋼.
　　Regional methods of low-temperature steel-refining.
　　Jiangsu Office of Metallurgical Industry, Nanjing, 1958.
　　Cf. Wagner (1985, pp. 60–66).

Anon. (*1958c*).
　　"精耕細作" 高產多收—"黃繼光爐" 的技術經驗.
　　Technical study of the Huang Jiguang blast furnace.
　　Yejin bao 冶金報 (Metallurgical news), 1958, no. 45, 20–22, 34.

Anon. (*1958d*).
　　Tufa yelian 土法冶煉.
　　Local methods of smelting [in Sichuan].
　　Chongqing Renmin, Chongqing, 1958.

Anon. (1959). *Das ganze Volk schmilzt Eisen und Stahl.* Verlag für Fremdsprachige Litteratur, Peking, 1959. Original, *All China makes iron and steel*, Peking 1958, not seen.

Anon. (1959b). 'L'industrie sidérurgique chinoise (1890–1959).' *Notes et études documentaires* (*La Documentation Française,* Paris), 12 nov. 1959 (no. 2591).

Anon. (*1960*).
萬福鋼鐵廠小高爐高產經驗.
High productivity with small blast furnaces at Wanfu Steelworks in Daxian 達縣, Sichuan.
Gangtie 鋼鐵 (Iron and steel), 1960, no. 6, 360.

Anon. (*1967*).
Jiao Fei zhan shi 剿匪戰史.
'A history of military actions against the Communist rebellion during 1930–1945'.
By the Military History Bureau, Ministry of National Defence 國防部史正局.
6 vols., Guofangbu Shizhengju & Zhonghua Dadian Bianyinhui 中華大典編印會, 1967.

Anon. (1979). *Direct reduction of iron ore: A bibliographical survey.* The Metals Society, London, 1979. Tr. of *Direktreduktion von Eisenerz: Eine bibliographische Studie*, Stahleisen, Düsseldorf, 1976

Anon. (*1981*).
Zhongguo shehui jingji shi luncong, di-yi-ji 中國社會經濟史論叢第一輯 .
Studies in Chinese social and economic history, collection 1.
Ed. by Shanxi Provincial Institute of Social Sciences 山西省社會科學研究所.
Shanxi Renmin Chubanshe, Taiyuan, 1981.

Anon. (*1983*).
Qing dai de kuangye 清代的礦業.
Sources on the mining industry in the Qing period.
2 vols., Zhonghua Shuju, 1983.

Anon. (*1985a*).
Shanxi jingji 山西經濟.
'Shanxi economy'.
Shanxi Renmin Chubanshe, Taiyuan, 1985.

Anon. (*1985b*).
Ming Qing Guangdong shehui jingji xingtai yanjiu 明清廣東社會形態研究.
Studies of the social–economic formation of Guangdong in the Ming and Qing periods.
Ed. by Guangdong historical society 廣東歷史學會.
Guangdong Renmin Chubanshe, Guangzhou, 1985.

Anon. (*1987a*).
Ming Qing Foshan beike wenxian jingji ziliao 明清佛山碑刻文獻經濟資料.
Source materials from stone inscriptions and documents for the economic

history of Foshan, Guangdong, in the Ming and Qing periods.
Comp. by Ancient Chinese History Seminar, Institute of History, Guangdong Provincial Academy of Social Sciences 廣東省社會科學院歷史研究所中國古代史研究室; Ancient Chinese History Seminar, Department of History, Sun Yat-sen University 中山大學歷史系中國古代史教研室; and Foshan, Guangdong, Municipal Museum 廣東省佛山市博物館.
Guangdong Renmin Chubanshe, 1987.

Anon. (*1987b*).
Ming Qing Guangdong shehui jingji yanjiu 明清廣東社會經濟研究.
Studies in the social and economic history of Guangdong in the Ming and Qing periods.
Guangdong Renmin Chubanshe, 1987.

Anon. (*1990*).
Yingshan xian zhi 應山縣志.
Gazetteer of Yingshan County, Hubei.
By an *ad hoc* editorial committee 湖北省應山縣志編纂委員會.
Hubei Kexue Chubanshe, 1990 (*Zhonghua Renmin Gongheguo Difangzhi congshu* 中華人民共和國地方志叢書).

Aoyama Sadao (*1933*) 青山定雄.
Dokushi hōyu kiyō sakuin: Shina rekidai chimei yōran 讀史方輿紀要索引支那歷代地名要覽.
Handbook of Chinese historical place-names: Index to the *Du shi fangyu ji-yao*.
Toho Bunka Hakuin 東方文化學院, Tokyo, 1933.

Ball, Samuel (1972). 'Observations on the expediency of opening a new port in China.' Murphey (1972, pp. 1–23). Orig. Foreign Office correspondence, FO 677/8, 1817.

Barraclough, Kenneth C. (1976). 'The development of the cementation process for the manufacture of steel.' *Post-medieval archaeology*, 1976, **10**, 65–88 + fold-out figures + plates 9–13.

Bays, Daniel H. (1970). 'The nature of provincial political authority in late Ch'ing times: Chang Chih-tung in Canton, 1884–1889.' *Modern Asian studies*, 1970, **4** (no. 4), 325–347.

Bielenstein, Hans (1987). 'Chinese historical demography, A.D. 2–1982.' *Bulletin of the Museum of Far Eastern Antiquities*, 1987, **59**, 1–288.

Biringuccio, Vannoccio (1942). *The Pirotechnia of Vannoccio Biringuccio*. Translated by M. T. Gnudi and Cyril Stanley Smith. American Institute of Mining Engineers, 1942. Orig. 1540. Tr. repr. 1959; facs. repr. M.I.T. Press, Cambridge, Mass. & London, 1966.

Boorman, Howard L. (1967–71). *Biographical dictionary of Republican China*. 4 vols., Columbia University Press, New York & London, 1967–71.

Bramall, Chris (1993). *In praise of Maoist economic planning: Living standards and economic development in Sichuan since 1931*. Clarendon Press, Oxford, 1993.

Bronson, Bennet (1987). Review of Wagner (1985). *Archeomaterials*, 1987, **2** (no. 1), 95–107.

Bronson, Bennett (1992). 'Patterns in the early Southeast Asian metals trade.' Glover et al. (1992, pp. 63–114).

Brown, Ian (ed.) (1989). *The economies of Africa and Asia in the inter-war depression*. Routledge, London & New York, 1989.

Brown, Shannon R. (1978). 'The partially opened door: Limitations on economic change in China in the 1860s.' *Modern Asian studies*, 1978, **12** (no. 2), 177–192.

Brown, Shannon R. (1979). 'The transfer of technology to China in the nineteenth century: The role of direct foreign investment.' *Journal of economic history*, 1979, **39** (no. 1), 181–197.

Cao Tengfei 曹騰騑 & Tan Dihua 譚棣華 (*1985*).
關於明清廣東冶鐵業的幾個問題.
Some problems concerning the iron industry in Guangdong in the Ming and Qing periods.
Anon. (*1985b*, pp. 117–143).

Cao Tengfei 槽騰騑 & Li Caiyao 李才垚 (*1985*).
廣東羅定古冶鐵爐遺址調查簡報.
Brief report on iron-smelting furnace sites in Luoding County, Guangdong.
Wenwu 文物 ('Cultural relics'), 1985, no. 12, 70–74.

Chaudhuri, K. N. (1985). *Trade and civilisation in the Indian Ocean: An economic history from the rise of Islam to 1750*. Cambridge University Press, Cambridge, 1985.

Chen Buwu (a.o.) (*1928*) 陳步武.
Xuxiu Dazhu xian zhi 續修大竹縣志.
Gazetteer of Dazhu County, Sichuan.
1928; facs. repr. with title *Dazhu xian zhi* 大竹縣志, 4 vols., Chengwen, Taibei, 1976 (*Zhongguo fangzhi congshu, Huazhong difang* 中國方志叢書華中地方, 380).

Chen Shantong 陳善同 et al. (*1936*).
Chongxiu Xinyang xian zhi 重修信陽縣志.
Revised gazetteer of Xinyang County, Henan.
Xinxing Yinshuguan, Hankou, 1936; facs. repr., 3 vols., Chengwen, Taibei, 1968 (*Zhongguo fangzhi congshu, Huabei difang* 中國方志叢書華北地方, 121).

Chen Shisong 陳世松 & Jia Daquan 賈大泉 (*1986*) (eds.).
Sichuan jianshi 四川簡史.
Brief history of Sichuan.
Sichuan Sheng Kexueyuan, Chengdu, 1986.

Cheong, W. E. (1965). 'Trade and finance in China: 1784–1834. A reappraisal.' *Business history*, 1965, **7** (no. 1), 34–56.

Clunas, Craig (1984). *Chinese export watercolours*. Victoria and Albert Museum, London, 1984.

Coggin Brown, J. (1920a). 'The mines and mineral resources of Yunnan, with short accounts of its agricultural products and trade.' *Memoirs of the Geological Survey of India*, 1920a, **47** (no. 1), 1–201 + plates 1–8.

Coggin Brown, J. (1920b). 'The mines and minerals of Yunnan, south China.' *The mining magazine*, 1920b, **23** (no. 5, 6), 267–277, 331–342. Note repr., *Far Eastern review*, 1921, **17** (nos. 4, 6, 7), 248–254, 391–395, 465–469.

Corbin, Paul L. (1913). 'The industrial future of Shansi province.' In George H. Blakeslee, ed., *Recent developments in China*, Stechert, New York, 1913; pp. 256–271.

Cordier, Henri (1902). 'Les marchands hanistes de Canton.' *T'oung pao, Sér. 2*, 1902, **3**, 281–315.

Cordier, Henri (1909). 'Catalogue des albums chinois et des ouvrages relatifs à la Chine conservées au Cabinet des Estampes de la Bibliothèque Nationale.' *Journal asiatique*, 1909, **Sér. 10, 14**, 209–262.

Courant, Maurice (1902–10). *Catalogue des livres chinois, coréens, japonais, etc. (Bibliothèque Nationale, Département des Manuscrits)*. 2 vols., Ernest Leroux, Paris, 1902–10.

Cremer, L. (1913). 'Bericht über eine Reise in der chinesischen Provinz Szetschuan.' *Zeitschrift für das Berg-, Hütten- und Salinenwesen im Preussischen Staate*, 1913, **61**, 1–146.

Crossman, Carl L. (1991). *The decorative arts of the China trade: Paintings, furnishings and exotic curiosities*. Antique Collectors' Club, Woodbridge, Suffolk, 1991. [Orig. *The China trade: Export painting, furniture, silver and other objects*, Princeton, 1972.]

[Daurand, Paul Émile] (1845). *La Chine ouverte: Aventures d'un fan-kouei dans le pays de Tsin*, par Old Nick. H. Fournier, Paris, 1845.

Dautremer, Joseph [1911]. *La grande artère de la Chine: Le Yangtseu*. Guilmoto, Paris, n.d. (*Librairie Orientale & Américaine*).

Davenport-Hines, R. P. T. & Geoffrey Jones (ed.) (1989). *British business in Asia since 1860*. Cambridge University Press, Cambridge, 1989.

Davidov, — (mining engineer) (1872a). 'O mineral'nyikh bogatstvakh Kul'dzhi i o sposobakh razrabotki ikh tuzemtsami' (On the mineral resources of Kul'dzhi and on the indigenous methods of exploiting them). *Gornyii Zhurnal (St. Petersburg)*, 1872 (no. 2), 193–212 + ill. Cf. 1872b.

Davidov, — (mining engineer) (1872b). 'Montanindustrie an der Grenze Chinas.' *Berg- und Hüttenmännischen Zeitung*, 1872b, **31**, 394–400 + Taf. 11. Tr. of 1872a by J. H. Langer.

Day, St. John V. (1875). *On the high antiquity of iron and steel*. W. H. Guest, London, 1875. 'Read before the Philosophical Society of Glasgow, April 28, 1875.'

Deng Kaisong (*1985*) 鄧開頌.
明至清代前期廣東鐵礦產地和冶爐分布的統計.
Statistics on the distribution of iron-ore production sites and smelting furnaces in Guangdong in the Ming and early Qing periods.
Anon. (*1985b*, pp. 170–186).

Dermigny, Louis (1964a). *Les mémoires de Charles de Constant sur le commerce à la Chine*. S.E.V.P.E.N, Paris, 1964a. (École Pratique des Hautes Études – VIe Section, Centre de Recherches Historiques: Ports – routes – trafics, 16).

Dermigny, Louis (1964b). *La Chine et l'Occident: Le commerce à Canton au XVIIIe siècle, 1719–1833*. 4 vols., S.E.V.P.E.N, Paris, 1964b. (École Pratique des Hautes Études – VIe Section, Centre de Recherches Historiques: Ports – routes – trafics, 18). 3 vols. + Album.

Di Xianghua (1987). 'Revisiting the Dabie Mountains.' *China reconstructs*, 1987 (no. 11), 19–23.

Dickmann, H. (1932). 'Primitive Verkokungs- und Eisendarstellungsverfahren in China.' *Beiträge zur Geschichte der Technik und Industrie* (Verein Deutscher Ingenieure, Berlin), 1932, **21**, 152–154.

Ding Richu & Shen Zuwei (1992). 'Foreign trade and China's economic modernisation.' Wright (1992, pp.165–176).

Ding Wenjiang (*1956*) 丁文江.
漫游散記.
'Miscellaneous field notes'.
Repr. *Zhongyang Yanjiuyuan yuankan* 中央研究院院刊 ('Annals of Academia Sinica', Taibei), **3**, 341–437.

Downing, C. Toogood (1838). *The Fan-qui in China in 1836–7*. Henry Colburn, London, 1838. Facs. repr. Irish University Press, Shannon, 1972.

Du shi fangyu jiyao 讀史方輿紀要.
Essentials of geography for reading history.
Comp. by Gu Zuyu 顧祖禹, 1631–92, before 1673. First publ. in the period 1796–1821.
Typeset repr., Erlin Zhai 二林齋, 1901.
Index Aoyama Sadao (*1933*). Cf. Hummel (1944, p. 420).

Du Shouhu 杜受祐 & Zhang Xuejun 張學君 (eds.) (*1987*).
Jinxiandai Sichuan changzhen jingji zhi (Di-er-ji) 近現代四川場鎮經濟志 （第二集）.
The economy of rural industry in modern and contemporary Sichuan (second collection).
Sichuan sheng Shehui Kexue Yuan Chubanshe 四川省社會科學院出版社, 1987.
Other collections not seen.

DuClos, P. (1898). 'Rapport sur les mines et la métallurgie.' In *La mission lyonnaise d'exploration commerciale en Chine, 1895–1897*, Chambre de Commerce de Lyon, Lyon, 1898; pp. 283–314.

Eames, James Bromley (1909). *The English in China: Being an account of the intercourse and relations between England and China from the year 1600 to the year 1843 and a summary of later developments*. 1909. Facs. repr. Curzon Press, London, 1974.

Eberstein, Bernd (1974). *Bergbau und Bergarbeiter zur Ming-Zeit (1368–1644)*. Gesellschaft für Natur- und Völkerkunde Ostasiens, Hamburg, 1974.

Fairbank, John King (1953). *Trade and diplomacy on the China coast: The opening of the treaty ports, 1842–1854*. Stanford University Press, Stanford, 1953. Orig. 2 vols.; 1 vol. ed., 1964; paperback ed. 1969.

Fan Baisheng (*1985*) 范百勝.
山西晉城坩堝煉鐵調查報告.
Report of an investigation (in the 1950s) of crucible smelting of iron in Jincheng, Shanxi.
Keji shi wenji 科技史文集 **13**, pp. 143–149.

Faure, David (1989). *The rural economy of pre-Liberation China: Trade expansion and peasant livelihood in Jiangsu and Guangdong, 1870 to 1937*. Oxford University Press, Hong Kong, 1989.

Faure, David (1990). 'What made Foshan a town? The evolution of rural-urban identities in Ming–Qing China.' *Late Imperial China*, 1990, **11** (no. 2), 1–31.

Feuillet de Conches, F. (1856). 'Les peintres européens en Chine et les peintres chinois.' *La revue contemporaine*, 1856, **25**, 216–260.

Feuillet de Conches, F. (1862). *Causeries d'un curieux: Varietés d'histoire er d'art tirées d'un cabinet d'autographes et de dessins*. Vol. 2. Henri Plon, Paris, 1862.

Foster, Frank A. (1926). 'Chinese make iron today as in the dim past.' *The foundry*, 1926, **54**, 173–177.

Gale, W. K. V. (1977). *Iron and steel*. Moorland, Buxton, 1977. (*Historic industrial scenes*).

Geerts, A. J. C. (1878–83). *Les produits de la nature japonaise et chinoise . . . Partie inorganique et minéralogique, contenant la description des minéraux et des substances qui dérivent du règne minérale*. 2 vols., C. Lévy, Yokohama, 1878–83. Vol. 1, 1878; vol. 2, 1883.

Glover, Ian, Pornchai Suchitta, & John Villiers (ed.) (1992). *Early metallurgy, trade and urban centres in Thailand and Southeast Asia*. White Lotus, Bangkok, 1992.

Goodman, David S. G. (1986). *Centre and province in the People's Republic of China: Sichuan and Guizhou, 1955–1965*. Cambridge University Press, Cambridge, 1986. (*Contemporary China Institute publications*).

Goodrich, L. Carrington & Chaoying Fang (1976). *Dictionary of Ming biography, 1368–1644*. 2 vols., Columbia University Press, New York & London, 1976.

Greenberg, Michael (1951). *British trade and the opening of China, 1800–42*. Cambridge University Press, Cambridge, 1951. Facs. repr. Monthly Review Press, London, n.d.

Guangdong xinyu 廣東新語.
New discourses on the province of Guangdong.
By Qu Dajun 屈大均, 1630–96.
Editions: (1) Shuidian'ge 水天閣, preface by Pan Lei 潘耒 dated 1700; facs. repr. Xuesheng shuju, Taibei, 1968. (2) Typeset repr., Zhonghua Shuju, 1974.

Hall, Bert S. (1983). 'Cast iron in late medieval Europe: A reexamination.' *CIM Bulletin*, 1983, **76** (no. 855), 86–90.

Hansen, Max (1958). *Constitution of binary alloys*. 2nd ed., McGraw–Hill, New York / Toronto / London, 1958. (Metallurgical and metallurgical engineering series).

Hao Yen-p'ing (1986). *The commercial revolution in nineteenth-century China: The rise of Sino–Western mercantile capitalism*. University of California Press, Berkeley, 1986.

Hara Zenshirō (*1991*) 原善四郎.
Chūgoku korai no seitetsusuhō to zuhō seitetsu: D B Wagunā sho 'Daibessan' no shōkai
中國古來の製鐵法と土法製鐵　ＤＢワグナー著「大別山」の紹介.
(Ancient Chinese iron production techniques and regional iron production: D. B. Wagner's *Dabieshan*). *Kinzoku* 金屬 (Metals), **61**.7: 61–69. Review of Wagner (1985).

Hara, Zenshiro (1992). 'Crucible smelting in Manchuria.' *Archeomaterials*, 1992, **6** (no. 2), 131–139.

Hara Zenshirō (*1993*) 原善四郎.
Chūgoku zuhō seitetsuhō no hitotsu: rutsubo seitetsuhō 中國土法製鐵法の１つ—ルツボ製鐵法.
One Chinese traditional iron-production method: crucible smelting.
Kinzoku 金屬 (Metals), 1993.4: 77–82.

Harley, C. Knick (1988). 'Ocean freight rates and productivity, 1740–1813: The primacy of mechanical invention reaffirmed.' *Journal of economic history*, 1988, **48** (no. 4), 851–876.

Hartwell, Robert M. (1963). *Iron and early industrialism in eleventh-century China*. Ph.D. dissertation, University of Chicago, 1963.

[Henderson, James] (1872?). *Notes of a walk through parts of Hupeh, Honan, Shansi, and Chih-li, in 1871*. [Shanghai?], [1872?]. The Wade Collection of the Cambridge University library has an imperfect copy of this exceedingly rare book. 'Repr. from the Shanghai Evening Courier and the Shanghai Budget, 1871–2.'

Herman, Theodore (1956). 'The role of cottage and small-scale industries in Asian economic development.' *Economic development and cultural change*, 1956, **5**, 356–370.

Hirth, Friedrich (1890). 'Die Handelsprodukte von Kuang-tung.' In his *Chinesische Studien*, G. Hirth's Verlag, München & Leipzig, 1890; pp. 76–101.

Hosie, Alexander (1901). *Manchuria: Its people, resources and recent history*. Methuen, London, 1901.

Hosie, Alexander (1922). *Szechwan: Its products, industries and resources*. Kelly & Walsh, Shanghai, 1922.

Hou Defeng 侯德封 & Cao Guoquan 曹國權 (*1946*).
三十年來中國之煤鑛事業.
China's coal-mining industry during the past thirty years.
Zhou Kaiqing (*1946*, pp. 815–839).

Hsiao Liang-lin (1974). *China's foreign trade statistics, 1864–1949*. East Asian Research Center, Harvard University, Cambridge, Mass., 1974.

Hu Boyuan (*1946*) 胡博淵.
三十年來中國之鋼鐵事業.
China's iron and steel industry during the past thirty years.
Zhou Kaiqing (*1946*, pp. 799–813).

Huang, C. T. (1919). 'Conversion of white iron into foundry: How Chinese native irons may be made available as a means of relieving the scarcity of other grades in that country.' *Iron age* (New York), 1919, **103** (no. 4), 231–232.

Huang Jianxin 黃建新 & Luo Yixing 羅一星 (*1987*).
論明清時期佛山城市經濟的發展.
The development of the urban economy of Foshan, Guangdong, in the Ming and Qing periods.
Anon. (*1987b*, pp. 26–56).

Huang Zhanyue 黃展岳 & Wang Daizhi 王代之 (*1962*).
雲南土法煉鐵的調查.
Investigation of local methods of iron smelting in Yunnan.
Kaogu 考古 ('Archaeology'), 1962, no. 7, 368–374 + 381.

Huard, P. & M. Wong (1966). 'Les enquêtes françaises sur la science et la technologie chinoises au XVIIIe siècle.' *BEFEO*, 1966, **52** (no. 1), 137–226.

Hucker, Charles O. (1985). *A dictionary of official titles in Imperial China*. Stanford University Press, Stanford, 1985.

Hummel, Arthur W. (1944). *Eminent Chinese of the Ch'ing period*. 2 vols., Library of Congress, Washington, D.C., 1944. Numerous reprints.

Imbault Huart, C. (1899). 'Le voyage de l'ambassade hollandaise de 1656 à travers la province de Canton.' *Journal of the China Branch of the Royal Asiatic Society, N.S.*, 1899, **30**, 1–73.

Ishikawa, Shigeru (1972). 'A note on the choice of technology in China.' *Journal of development studies*, 1972, **9** (no. 1), 161–186.

Jack, R. Logan (1904). *The back blocks of China: A narrative of experiences among the Chinese, Sifans, Lolos, Tibetans, Shans and Kachins, between Shanghai and the Irrawadi*. Edward Arnold, London, 1904.

Jiang Tianfeng (ed.) (*1992*) 江天鳳.
Changjiang hangyun shi (jindai bufen) 長江航運史（近代部分）.
A history of navigation on the Yangzi (modern times).
Renmin Jiaotong Chubanshe 人民交通出版社, Beijing, 1992.

Jiang Zuyuan (*1987a*) 蔣祖緣.
清代佛山商人的構成及其對商業的影響.
The organisations of merchants in Foshan, Guangdong, in the Qing period and their effect on commerce.
Guangzhou yanjiu 廣州研究 (Research on Guangzhou), 1987, no. 8, 54–57.

Jiang Zuyuan (*1987b*) 蔣祖緣.
   清代佛山的商業和商人.
   Commerce and merchants in Foshan, Guangdong, in the Qing period.
   Anon. (*1987b*, pp. 1–25).

Johannsen, Otto (1911–17). 'Die Quellen zur Geschichte des Eisengusses im Mittel-
   alter und in der neueren Zeit bis zum Jahre 1530.' *Archiv für die Geschichte
   der Naturwissenschaften und der Technik* (Leipzig), 1911, **3**, 365–394; 1914,
   5, 127–141; 1917, **8**, 66–81.

Johannsen, Otto (1941). 'Die Erzeugung von flüssigem Roheisen im Hochofen.'
   Technikgeschichte, 1941, **30**, 57–62 + Tafel 32.

Kapp, Robert A. (1973). *Szechwan and the Chinese Republic: Provincial milita-
   rism and central power, 1911–1938*. Yale University Press, New Haven &
   London, 1973.

Kocher, E. (1921). 'Die neuere Entwicklung der chinesischen Eisenindustrie.'
   Stahl und Eisen, 1921, **41** (no. 1), 9–12.

Kong Lingtan (*1957*) 孔令壇.
   山西省的兩種土法煉鐵.
   Introducing two traditional methods of iron smelting in Shanxi.
   Gangtie 鋼鐵 (Iron and steel), 1957, no. 59, 85–89.

Kreitner, Gustav (1881). *Im fernen Osten: Reisen des Grafen Béla Széchenyi in In-
   dien, Japan, China, Tibet und Birma in den Jahren 1877–1880*. Alfred
   Hölder, Wien, 1881.

Lavollée, M. C. (1853). *Voyage en Chine: Ténériffe – Rio-Janeiro – Le Cap – île
   Bourbon – Malacca – Singapore – Manille – Macao – Canton – Ports chinois
   – Cochinchine – Java*. Just Rouvier & A. Ledoyen, Paris, 1853.

Le Fevour, Edward (1968). *Western enterprise in late Ch'ing China: A selective
   survey of Jardine, Matheson and Company's operations, 1842–1895*. East
   Asian Research Center, Harvard University, Cambridge, Mass., 1968. (*Harvard East Asian monographs*, 26).

Lee, Robert (1989). *France and the exploitation of China, 1885–1901: A study in
   economic imperialism*. Oxford University Press, Hong Kong, 1989.

Legge, James (tr.) (1872). *The Chinese classics: With a translation, critical and ex-
   egetical notes, prolegomena, and copious indexes*. Vol. 5, pts. 1–2: *The
   Ch'un ts'ew, with the Tso chuen*. Lane, Crawford, Hongkong, 1872. Facs.
   repr. Hong Kong University Press, 1960. Numerous later reprints.

Lei Baohua (*1943*) 雷寶華.
   抗戰以來四川之礦業.
   The mining industry in Sichuan after the War.
   Sichuan jingji jikan 四川經濟季刊 (Sichuan economic quarterly), 1943, **1**
   (no. 1), 44–47.

Leonard, Jane Kate (1984). *Wei Yuan and China's rediscovery of the maritime
   world*. Council on East Asian studies, Harvard university, Cambridge, Mass.,
   1984. [Wei Yuan 魏源, 1794–1857].

Li Longqian (*1981*) 李龍潛.
清代前期廣東採礦、冶鑄業中的資本主義萌芽.
Embryonic capitalism in the mining and metallurgical industry of Guangdong in the early Qing period.
Anon. (*1981*), pp. 352–380.

Li Renkuan (*1959*) 李仁寬.
四川合川鋼鐵廠和江北鋼鐵廠小高爐長壽的經驗介紹.
Experience with long-lived small blast furnaces at Hechuan Steelworks and Jiangbei Steelworks in Sichuan.
Gangtie 鋼鐵 (Iron and steel), 1959, no. 6, 198–199.

Li Runtian (ed.) (*1987*) 李潤田.
Henan sheng jingji dili 河南省經濟地理.
Economic geography of Henan province.
Xinhua Chubanshe, Beijing, 1987.

Liang Guang yan fa zhi 兩廣鹽法志.
The administration of the salt laws in Guangdong and Guangxi.
Comp. by Li Shiyao 李侍堯, Tuo Enduo 託恩多, and Su Chang 蘇昌.
Publ. by the Ministry of Revenue (*Hu bu* 戶部), 1762.
Rare books collection, School of Oriental and African Studies, London. 'Deposited on loan by University College, London.'

Licent, Émile (1924). *Dix années (1914–1923) dans le bassin du Fleuve Jaune et autres tributaires du Golfe du Pei Tcheu Ly*. Vol. 2. Librairie Française / Imprimerie de la Mission Catholique Sienhien, Tientsinn, 1924.

Liljevalch, C. F. (1848). *Chinas handel, industri och statsförfattning, jemte underrättelser om chinesernes folkbildning, seder och bruk, samt notiser om Japan, Siam m. fl. orter*. Beckman, Stockholm, 1848.

Liu Jixian 劉集賢, Kong Fanzhu 孔繁珠, & Wan Liangshi 萬良適 (*1982*).
Shanxi ming chan 山西名產.
Famous products of Shanxi.
Shanxi Renmin Chubanshe, Taiyuan, 1982.

Liu, Ta-chung & Kung-chia Yeh (1965). *The economy of the Chinese mainland: National income and economic development, 1933–1959*. Princeton University Press, Princeton, 1965.

Liu Xu (*1989*) 劉旭.
Zhongguo gudai huopao shi 中國古代火炮史.
The ancient history of cannons in China.
Shanghai Renmin, Shanghai, 1989.

Liu Zhengtan 劉正埮, et al. (*1984*).
Hanyu wailaici cidian 漢語外來詞詞典.
Dictionary of foreign loan-words in Chinese.
Shanghai Cishu Chubanshe, Shanghai, 1984.

Liu Zhichao 劉志超 & Tang Youyu 唐有餘 (*1959*).
湖北省麻城縣黃繼光爐高產經驗.
High production with the Huang Jiguang blast furnace in Macheng County, Hubei.
Gangtie 鋼鐵 (Iron and steel), 1959, no. 6, pp. 182–187.

Liu Zhiwei (*1986*) 劉志偉.
明清時期廣州城市經濟的特色.
Characteristics of the urban economy of Guangzhou in the Ming and Qing periods.
Guangzhou yanjiu 廣州研究 (Research on Guangzhou), 1986, no. 1, 62–65.

Lóczy, Lajos (1923). *Gróf Széchenyi Béla Emlékezete.* Magyar Tudományos Akadémia, Budapest, 1923. (*A Magyar Tudományos Akadémia Elhunyt tagjai felett tartott emlékbeszédek*, 18.8). Obituary of Count Béla Széchenyi, with bibliography.

Lu Manping 陸滿平 & Jia Xiuyan 賈秀岩 (*1992*).
Minguo jiage shi 民國價格史.
Price history in the Republican period.
Zhongguo Wujia Chubanshe 中國物價出版社, n.p., 1992.

Lu'an zhou zhi 六安州志.
Gazetteer of Lu'an Subprefecture, Anhui.
Comp. by Liu Gai 劉垓 et al., 1584.
Repr., 2 vols., Chengwen, Taibei, 1983 (*Zhongguo fangzhi congshu, Huazhong difang* 中國方志叢書華中地方, 615).

Luo Hongxing (*1983*) 羅紅星.
明至清前期佛山冶鐵業初探.
Preliminary study of the iron industry of Foshan, Guangdong, in the Ming and early Qing periods.
Zhongguo shehui jingji shi yanjiu 中國社會經濟史研究 ('Journal of Chinese social and economic history'), 1983, no. 4, 44–54.

Luo Mian (*1936*) 羅冕.
Gang tie 鋼鐵.
Iron and steel.
Report no. 15 in *Zhongguo Gongchengshi Xuehui Sichuan Kaochatuan baogao* 中國工程師學會四川考查團報告 (Reports of the Sichuan Investigation Group, Chinese Engineering Society).
N.p., n.d., preface dated 1936.

Luo Yixing (*1984*) 羅一星.
關於明清 "佛山鐵廠" 的幾點質疑.
Doubts concerning certain 'Foshan ironworks'.
Xueshu yanjiu 學術研究 ('Journal of academic research'), 1984, no. 1, 109–112.

Luo Yixing (*1985*) 羅一星.
明清時期佛山冶鐵業研究.
The iron industry of Foshan, Guangdong, in the Ming and Qing periods.
Anon. (*1985b*, pp. 75–116).

Lux, Fr. (1912). 'Koksherstellung und Hochofenbetrieb im Innern Chinas.' *Stahl und Eisen*, 1912, **22** (no. 34), 1404–1407.

MacFarquhar, Roderick (1983). *The origins of the Cultural Revolution.* Vol. 2: *The Great Leap Forward.* Oxford University Press, Oxford, 1983.

Magnusson, Gert (1985). 'Lapphyttan – an example of medieval iron production.' In *Medieval iron in society: Papers presented at the symposium in Norberg, May 6–10, 1985*, Jernkontoret & Riksantikvarieämbetet, Stockholm, 1985; pp. 21–57.

Mao Zedong 毛澤東 (1980). *Mao Zedong and the political economy of the Border Region*. Translated by Andrew Watson. Cambridge University Press, Cambridge, 1980. Orig. *Kang-Ri shiqi de jingji wenti yu caizheng wenti* 抗日時期的經濟問題與財政問題, December 1942.

Mao Zedong 毛澤東 (1990). *Report from Xunwu*. Translated by Roger R. Thompson. Stanford University Press, Stanford, Calif., 1990. Orig. *Xunwu diaocha* 尋烏調查, 1937, first publ. 1982.

McColl, Robert W. (1967). 'The Oyüwan Soviet area, 1927–1932.' *Journal of Asian studies*, 1967, **27** (no. 1), 41–60.

Meng Xianzhang (*1943*) 孟憲章.
四川在中國經濟史上的地位.
The place of Sichuan in Chinese economic history.
Sichuan jingji jikan 四川經濟季刊 (Sichuan economic quarterly), 1943, **1** (no. 1), 186–189.

Miao Changxing 苗長興 & Li Jinghua 李京華 (1994). 'Placer iron smelting in Xinyang, Henan Province.' In *BUMA-3: The Third International Conference on the Beginning of the Use of Metals and Alloys, 18–23 April 1994*, Sanmenxia, 1994; p. 91.

Mitchell, B. R. (1971). *Abstract of British historical statistics*. Cambridge University Press, Cambridge, 1971.

Moore-Bennet, Arthur J. (1915). 'The mineral area of western China.' *Far Eastern Review*, 1915, **12** (no. 6), 215–227.

Morse, H. B. (1922). 'The provision of funds for the East India Company's trade at Canton during the eighteenth century.' *Journal of the Royal Asiatic Society*, 1922, 227–255.

Morse, Hosea Ballou (1926–29). *The chronicles of the East India Company trading to China 1635–1834*. 5 vols., Clarendon Press, Oxford, 1926–29.

Mott, R. A. & Peter Singer (1983). *Henry Cort: The great finer. Creator of puddled iron*. The Metals Society, London, 1983.

Moulder, Frances V. (1977). *Japan, China, and the modern world economy: Toward a reinterpretation of East Asian development ca. 1600 to ca. 1918*. Cambridge University Press, Cambridge, 1977.

Muan, Arnulf & E. F. Osborn (1965). *Phase equilibria among oxides in steelmaking*. Pergamon Press, Oxford, 1965.

Murphey, Rhoads (ed.) (1972). *Nineteenth century China: Five imperialist perspectives*. Ann Arbor, Mich., Center for Chinese Studies, University of Michigan, 1972. (*Michigan papers in Chinese studies*, 13.)

Myers, Ramon H. (1989). 'The world depression and the Chinese economy 1930–6.' Brown (1989, pp. 253–278).

Nan Yue biji 南越筆記.
> Notes on travels in Guangdong.
> Li Tiaoyuan 李調元 (1734–1803), 1780.
> *Han hai* 函海, repr. Ledao Zhai 樂道齋, 1881–82; facs. repr., Hongye Shuju 宏業書局, Taibei, 1968.
> See Hummel (1944, pp. 486–488).

Needham, Joseph (1958). *The development of iron and steel technology in China.* The Newcomen Society, London, 1958. Second Dickinson Memorial Lecture to the Newcomen Society, 1956.

Needham, Joseph & Dorothy Needham (ed.) (1948). *Science outpost: Papers of the Sino–British Science Co-Operation Office (British Council Scientific Office in China) 1942–46.* London, Pilot Press, 1948.

Nyström, Erik T. (1910). 'Kinas skatkammare.' In Chr. Barthel, ed., *Festskrift tillägnad Peter Klason på hans sextioårsdag av lärjungar*, Stockholm, 1910; pp. 391–406.

Nyström, Erik T. (1912). *The coal and mineral resources of Shansi province, China, analytically examined.* Norstedt, Stockholm, 1912.

Osterhammel, Jürgen (1989). 'British business in China, 1860s – 1950s.' Davenport–Hines & Jones (1989, pp. 189–216).

Ou Chu (*1986*) 歐初.
> 《屈大均全集》序.
> Preface to *Complete works of Qu Dajun* (1630–1696).
> *Guangzhou yanjiu* 廣州研究 (Research on Guangzhou), 1986, no. 8, 49–52.

Paulinyi, Akos (1987). *Das Puddeln: Ein Kapitel aus der Geschichte des Eisens in der Industriellen Revolution.* Oldenbourg, München, 1987. (*Deutsches Museum von Meisterwerken der Naturwissenschaft und Technik: Abhandlungen und Berichte*, N.F., Bd. 4).

Peacey, J. G. & W. G. Davenport (1979). *The iron blast furnace: Theory and practice.* Pergamon Press, Oxford, 1979. (*International series on materials science and technology*).

Peng Jianxin (*1994*) 彭建新.
> 廣東省１９５８年大煉鋼鐵的情況分.
> The circumstances and consequences of the 1958 Campaign for Iron and Steel Production in Guangdong.
> *Guangdong shizhi* 廣東史志 (Guangdong historical treatises), 1994, no. 2, pp. 39–44.

Peng Zeyi (ed.) (*1957*) 彭澤益.
> *Zhongguo jindai shougongye shi ziliao (1840–1949)* 中國近代手工業史資料〔１８４０‐１９４９〕.
> Source materials on handicraft industry in modern China (1840–1949).
> 4 vols., Sanlian Shudian, Beijing, 1957 (*Zhongguo Kexueyuan Jingji Yanjiusuo Zhongguo jindai jingjishi cankao ziliao congkan* 中國科學院經濟研究所中國近代經濟參考資料叢刊, 4).

Percy, John (1861). *Metallurgy: The art of extracting metals from their ores, and adapting them to various purposes of manufacture.* [Vol. 1:] *Fuel; fire-clays; copper; zinc; brass; etc.* John Murray, London, 1861. Facs. repr. in 2 pts., Eindhoven: De Archaeologische Pers Nederland, n.d. [ca. 1985].

Percy, John (1864). *Metallurgy . . .* [Vol. 2:] *Iron; steel.* John Murray, London, 1864. Facs. repr. in 3 pts., Eindhoven: De Archaeologische Pers Nederland, n.d. [ca. 1983].

Pu Xiaorong (ed.) (*1993*) 蒲孝榮.
Zhonghua Renmin Gongheguo diming cidian: Sichuan sheng 中華人民共和國地名詞典　四川省.
Dictionary of place-names of the People's Republic of China: Szechwan Province.
Shangwu Yinshuguan, Beijing, 1993.

Qiao Zhiqiang (*1978*) 喬志強.
Shanxi zhitie shi 山西製鐵史.
The history of ironmaking in Shanxi.
Shanxi Renmin Chubanshe, Taiyuan, 1978.

Qin Dazhang 秦達章 & He Guoyou 何國佑 (*1905*).
Huoshan xian zhi 霍山縣志.
Gazetteer of Huoshan County, Anhui.
1905 ed., facs. repr. 3 vols., Chengwen, Taibei, 1974 (*Zhongguo fangzhi congshu, Huazhong difang* 中國方志叢書華中地方, 226).

Rawski, Thomas G. (1989). *Economic growth in prewar China.* University of California Press, Berkeley, 1989.

Read, Thomas T. (1911). 'The mineral production and resources of China.' *Transactions of the American Institute of Mining Engineers*, 1911, **43**, 3–53.

Read, Thomas T. (1921). 'Primitive iron smelting in China.' *Iron age*, 1921, **108** (no. 8), 451–455.

Read, Thomas T. (1937). 'Chinese iron – a puzzle.' *Harvard Journal of Asiatic Studies*, 1937, **2**, 398–407.

Reardon–Anderson, James (1991). *The study of change: Chemistry in China, 1840–1949.* Cambridge University Press, Cambridge, 1991.

Richard, L. (1908). *Comprehensive geography of the Chinese Empire and dependencies.* Translated by S. Kennelly. T'usewei Press, Shanghai, 1908. Orig. *Géographie de l'Empire de Chine*, 1905.

Richardson, H. L. (1945). 'Szechwan during the war.' *The geographical journal*, 1945, **106** (no. 1/2), 1–25 + plates.

von Richthofen, Ferdinand [1872?]. *Baron Richthofen's letters, 1870–1872 [to the Shanghai Chamber of Commerce].* North China Herald, Shanghai, n.d.

von Richthofen, Ferdinand (1877–1912). *China: Ergebnisse eigener Reisen und darauf gegründeter Studien.* 5 vols., Dietrich Reimer, Berlin, 1877–1912. Bd. 1: *Einleitender Theil*, 1877. Bd. 2: *Das nördliche China*, 1882. Bd. 3: *Das südliche China*, nach den hinterlassenen Manuscripten im letztwilligen

Auftrag des Verfassers hrsg. v. Ernst Tiessen, 1912. Bd. 4–5: *Palaeontologischer Theil* . . . , 1883–1911.

von Richthofen, Ferdinand (1907). *Ferdinand von Richthofen's Tagebücher aus China*. 2 vols., Dietrich Reimer, Berlin, 1907. Ausgew. u. hrsg. v. E. Tiessen.

Robertson, J. A. T. (1916). 'An engineer's travels in western China.' *The mining magazine*, 1916, **15** (no. 5), 267–284.

Rocher, Émile (1879–80). *La province chinoise du Yün-nan*. 2 vols., Leroux, Paris, 1879–80.

Rosenholtz, Joseph L. & Joseph F. Oesterle (1938). *The elements of ferrous metallurgy*. 2nd ed., Wiley / Chapman & Hall, New York / London, 1938.

Rosenqvist, Terkel (1974). *Principles of extractive metallurgy*. McGraw–Hill, New York, 1974. (*McGraw–Hill series in materials science and engineering*).

Rostoker, William (1987). Review of Wagner (1985). *Technology and culture*, 1987, **28** (no. 2), 346–347.

Rostoker, William & Bennet Bronson (1990). *Pre-industrial iron: Its technology and ethnology*. Privately published, Philadelphia, 1990. (*Archeomaterials monograph*, no. 1).

Rowe, William T. (1984). *Hankow: Commerce and society in a Chinese city, 1796–1889*. Stanford University Press, Stanford, Calif., 1984.

Shisan jing zhushu 十三經注疏
    The Thirteen Classics, with collected commentaries and sub-commentaries. Song. Ed. of Ruan Yuan 阮元, 1816; repr. Shijie Shuju, Shanghai, 1935.

Shockley, William H. (1904). 'Notes on the coal- and iron-fields of southeastern Shansi, China.' *Transactions of the American Institute of Mining Engineers*, 1904, **34**, 841–871.

Sidney, L. P. (1920). 'Blast furnace practice.' *The Times engineering supplement*, April 1920 (no. 546), 129; June 1920 (no. 548), 189.

Sieurin, Emil (1911). 'Höganäs järnsvamp.' *Jernkontorets annaler*, 1911, 448–493.

Skinner, G. William (1987). 'Sichuan's population in the nineteenth century: Lessons from disaggregated data.' *Late Imperial China*, 1987, **8** (no. 1), 1–99. Note that this printing of the article replaces a faulty printing in the previous issue of the same journal, 1986, **7** (no. 2), 1–79.

Smith, Paul J. (1988). 'Commerce, agriculture, and core formation in the upper Yangzi, 2 A.D. to 1948.' *Late Imperial China*, 1988, **9** (no. 1), 1–78.

Spence, Jonathan D. (1990). *The search for modern China*. Hutchinson, London, 1990.

Swamy, Subramanian (1973). *Economic growth in China and India 1952–1970*. University of Chicago Press, Chicago & London, 1973.

Széchenyi, Béla (1890). *Keletázsiai Utjának: Tudományos Eredménye 1977–1880. Elsö kötet: Az utazáson tett észlelések*. Kilián Frigyes Egyetemi Könyvárus Bizományában, Budapest, 1890. Cf. 1893.

Széchenyi, Béla (1893). *Die wissenschaftlichen Ergebnisse der Reise des Grafen Béla Széchenyi in Ostasien,* Bd. 1: *Die Beobachtungen während der Reise.* Hölzel, Wien, 1893. Translation of 1890.

Tan Dihua (*1987*) 譚棣華.
從《佛山街略》看明清時期佛山工商業的發展.
The development of commerce and industry in Foshan, Guangdong, in the light of the late Qing text, 'Foshan street guide'.
Qing shi yanjiu tongxun 清史研究通訊 (Bulletin of Qing historical research), 1987, no. 1, 6–14.

Tan Kesheng 譚克繩 & Ouyang Zhiliang 歐陽植梁 (*1987*).
E–Yu–Wan geming genjudi douzheng shi jianbian 鄂豫皖革命根據地鬥爭史簡編.
A short history of the struggle for the E–Yu–Wan Revolutionary Base Area. Jiefangjun Chubanshe, Beijing, 1987.

Tegengren, F. R. (1923–24). *The iron ores and iron industry of China: Including a summary of the iron situation of the circum-Pacific region.* 2 vols., Geological Survey of China, Ministry of Agriculture and Commerce, Peking (*Memoirs of the Geological Survey of China, series A,* no. 2). Part I 1921–23, part II 1923–24. English text and abridged Chinese translation by Xie Jiarong 謝家榮. Chinese title: *Zhongguo tiekuang zhi* 中國鐵礦誌, by Ding Gelan 丁格蘭 (*Dizhi zhuanbao,* A.2 地質專報甲種第二號). Xerographic repr. available from University Microfilms, Ann Arbor, Mich., order no. OP 12715.

Turner, Thomas (1895). *The metallurgy of iron and steel.* Vol. 1: *The metallurgy of iron.* Charles Griffin, London, 1895.

Tylecote, R. F. (1987). *The early history of metallurgy in Europe.* Longman, London & New York, 1987. (*Longman archaeology series*).

Viraphol, Sarasin (1977). *Tribute and profit: Sino–Siamese trade, 1652–1853.* Council on East Asian Studies, Harvard University, Cambridge, Mass., 1977.

Wagner, Donald B. (1985). *Dabieshan: Traditional Chinese iron-production techniques practised in southern Henan in the twentieth century.* Curzon Press, London & Malmö, 1985. (*Scandinavian Institute of Asian Studies Monograph series,* no. 52).

Wagner, Donald B. (1993). *Iron and steel in ancient China.* Brill, Leiden, 1993. (*Handbuch der Orientalistik, vierte Abteilung: China,* no. 9).

Wakeman, Frederic (1966). *Strangers at the gate: Social disorder in south China, 1839–1861.* University of California Press, Berkeley & Los Angeles, 1966.

Wang Hongjun 王宏鈞 & Liu Ruzhong 劉如仲 (*1980*).
廣東佛山資本主義萌芽的幾點探討.
Points for discussion concerning embryonic capitalism in Foshan, Guangdong.
Zhongguo Lishi Bowuguan guankan 中國歷史博物館館刊 ('Bulletin of the Museum of Chinese History'), 1980, **2**, 58–79.

Wang Jingzun 王景尊 & Wang Yuelun 王曰倫 (*1930*).
    正太鐵路 線地質礦產.
    C. T. Wang & Y. L. Wang, 'A study of the general and economic geology
    along the Chêng–T'ai (Shansi) Railway'. Chinese and English text.
    *Geological bulletin / Bulletin of the Geological Survey of China / Dizhi hui-
    bao* 地質 彙報, **15**, Engl. text pp. 53–118 + Ch. text pp. 49–92 + plates +
    maps.

Wang Ziyou (*1940*) 王子祐.
    小規模煉鐵爐設計問題之討論.
    The design of small-scale iron-smelting furnaces.
    Kuangye banyuekan 礦業半月刊 (Mining semimonthly, Chongqing), 1940,
    3 (no. 11/12/13/14), 1–15.

Way, Herbert W. L. (1916). 'The minerals of Sze-chuan, China.' *The mining mag-
    azine*, 1916, **15** (no. 1), 20–23.

Wiens, Herold J. (1949). 'The Shu Tao or road to Szechwan.' *The geographical
    review*, 1949, **39** (no. 4), 584–604.

Williamson, Alexander (1870). *Journeys in north China, Manchuria, and eastern
    Mongolia; with some account of Corea*. 2 vols., Smith, Elder & Co., London,
    1870.

Wright, Tim (ed.) (1992). *The Chinese economy in the early twentieth century: Re-
    cent Chinese studies*. St. Martin's Press, New York, 1992. Orig. Macmillan,
    London, 1992.

Wu Lansheng 吳蘭生, Liu Tingfeng 劉廷鳳, et al. (*1920*).
    Qianshan xian zhi 潛山縣志.
    Gazetteer of Qianshan County.
    1929 ed., repr., 2 vols., Chengwen, Taibei, 1985 (*Zhongguo fangzhi congshu,
    Huazhong difang* 中國方志叢書華中地方, 709).

Wu Yuwen (ed.) (*1985*) 吳郁文.
    Guangdong sheng jingji dili 廣東省經濟地理.
    Economic geography of Guangdong province.
    Xinhua Chubanshe, Beijing, 1985.

Xia Xiangrong 夏湘蓉, Li Zhongjun 李仲均, & Wang Genyuan 王根元 (*1980*).
    Zhongguo gudai kuangye kaifa shi 中國古代礦業開發史.
    The history of the development of the mining industry in ancient China.
    Dizhi Chubanshe 地質出版社, Beijing, 1980.

Xian Jianmin (*1993*) 冼劍民.
    清代前期廣東手工業的發展及其特點.
    The development of handicraft industry in Guangdong in the early Qing pe-
    riod and its characteristics.
    Guangdong shehui kexue 廣東社會科學 ('Social sciences in Guangdong'),
    1993, no. 4, 70–77.

Xu Jin 徐錦, Hu Jianying 胡鑑瑩, et al. (*1920*).
    Yingshan xian zhi 英山縣志.
    Gazetteer of Yingshan County, Anhui.

Maoqingyun Tang 毛青雲堂, 1920; repr., 4 vols., Chengwen, Taibei, 1985 (*Zhongguo fangzhi congshu, Huazhong difang* 中國方志叢書華中地方, 660).

Xu Junming (*1956*) 徐俊鳴.
Liang Guang dili 兩廣地理.
Geography of Guangdong and Guangxi.
Xin Zhishi Chubanshe 新知識出版社, Shanghai, 1956.
Xerox repr., Center for Chinese Research Materials, Washington, D.C., n.d.

Yan Zhaoping (*1936*) 晏兆平.
Guangshan xian zhi yuegao 光山縣志約稿.
Draft gazetteer of Guangshan County, Henan.
1936 ed., facs. repr. 2 vols., Chengwen, Taibei, 1968 (*Zhongguo fangzhi congshu, Huabei difang* 中國方志叢書華北地方, 125).

Yang Dajin (ed.) (*1938*) 楊大金.
Xiandai Zhongguo shiye zhi 現代中國實業誌.
Gazetteer of modern Chinese industry and commerce.
Changsha, 1938; facs. repr. in 3 vols., Huashi Chubanshe 華世出版社, Taibei, 1978 (*Zhongguo jingjishi shiliao congshu* 中國經濟史料叢書, E0105).

Yang Kuan (*1960*) 楊寬.
Zhongguo tufa yetie lian'gang jishu fazhan jianshi 中國土法冶鐵煉鋼技術發展簡史.
A brief history of the development of Chinese regional siderurgical techniques.
Renmin Chubanshe, Shanghai, 1960.

Yang Kuan (*1982*) 楊寬.
Zhongguo gudai yetie jishu fazhan shi 中國古代冶鐵技術發展史.
The history of the development of siderurgical technology in ancient China.
Renmin Chubanshe, Shanghai, 1982.

Ye Xian'en (*1987*) 葉顯恩.
清代廣東水運與社會經濟.
Social and economic aspects of water transport in Guangdong in the Qing period.
Zhongguo shehui jingji shi yanjiu 中國社會經濟史研究 ('Journal of Chinese social and economic history'), 1987, no. 4, 1–10.

Ye Xian'en & Chen Chunsheng (1990). 'Social and economic history of Guangdong province: State of the field.' *Late Imperial China*, 1990, **11** (no. 2), 102–115. Rev. & tr. by Robert Y. Eng.

Yu Jinfang 余晉芳 (*1935a*).
Macheng xian zhi qianbian 麻城縣志前編.
Gazetteer of Macheng County, Hubei, first part.
Zhong–Ya Yinshuguan 中亞印書館, Hankou, 1935; facs. repr. 4 vols., Chengwen, Taibei, 1975 (*Zhongguo fangzhi congshu, Huazhong difang* 中國方志叢書華中地方, 357).

Yu Jinfang 余晉芳 (*1935b*).
Macheng xian zhi xubian 麻城縣志續編.
Gazetteer of Macheng County, Hubei, continuation.
Zhong–Ya Yinshuguan 中亞印書館, Hankou, 1935; facs. repr. 4 vols.,
Chengwen, Taibei, 1975 (*Zhongguo fangzhi congshu, Huazhong difang* 中國
方志叢書華中地方, 357).

Yu Qinglan 俞慶瀾, Zhang Cankui 張燦奎, et al. (*1921*).
Susong xian zhi 宿松縣志.
Gazetteer of Susong County, Anhui.
1921; repr. Chengwen, Taibei, 1984 (*Zhongguo fangzhi congshu, Huazhong
difang* 中國方志叢書華中地方, 671).

Yu Siwei (*1983*) 余思偉.
清代前期廣州與東南亞的貿易關系.
Trade between Guangzhou and Southeast Asia in the early Qing period.
Zhongshan Daxue xuebao (Zhexue shehui kexue ban) 中山大學學報（哲學
社會科學版□ ('Journal of Sun Yatsen University: Social sciences edition'),
1983, no. 2, 73–83.

Yue dong cheng'an chubian 粵東成案初編.
Guangdong legal cases, first collection.
Comp. by Zhu Yun 朱橒, preface dated 1828. 'New printing', 1832.
Wade Collection, Cambridge University Library.

Yue dong wenjian lu 粵東聞見錄.
Record of things seen and heard in Guangdong.
By Zhang Ju 張渠, d. +1740.
Typeset ed., bound with *Nan Yue youji* 南越遊記, Guangdong Gaodeng
Jiaoyu Chubanshe 廣東高等教育出版社, 1990.

Yue zhong jian wen 粵中見聞.
Things seen and heard in Guangdong.
By Fan Tuanang 范端昂, eighteenth cent.
Typeset ed., ed. by Tang Zhiyue 湯志岳, Guangdong Gaodeng Jiaoyu Chu-
banshe 廣東高等教育出版社, 1988.

Zhang Chengji (*1959*) 張成吉.
四川合川鋼鐵廠爐外去硫試驗報告.
Experiments with the removal of sulphur outside the furnace at Hechuan
Steelworks in Sichuan.
Gangtie 鋼鐵 (Iron and steel), 1959, no. 6, 199–203.

Zhang Xiaomei (ed.) (*1939*) 張肖梅.
Sichuan jingji cankao ziliao 四川經濟參考資料.
Reference materials on Sichuan's economy.
Zhongguo Guomin Jingji Yanjiusuo 中國國民經濟研究所, Shanghai, 1939.

Zhang Xuejun 張學君 & Zhang Lihong 張莉紅 (*1990*).
Sichuan jindai gongye shi 四川近代工業史.
The history of modern industry in Sichuan.
Sichuan Renmin Chubanshe, Chengdu, 1990.

Zhang Youxian 張友賢 & Guo Yujing 郭玉璟 (*1932*).
河南鐵礦.
Iron ores of Henan province.
Henan sheng dizhi diaochasuo huikan 河南省地質調查所彙刊, **1**, pp. 221–243.

Zhang Zhidong (*1937*) 張之洞.
Zhang Wenxiang gong quanji 張文襄公全集.
Complete works of Zhang Zhidong (1837–1909), comp. by Wang Shudan 王樹枏.
Chuxue Jinglu 楚學精廬, Beiping, 1937; facs. repr., 11 vols., Wenhai Chubanshe 文海出版社, Taibei, 1970.

Zhou Kaiqing (ed.) (*1946*) 周開慶.
Sanshi nian lai zhi Zhongguo gongcheng: Zhongguo Gongchengshi Xuehui sanshi zhounian jiniankan 三十年來之中國工程　中國工程師學會三十週年紀年刊.
Chinese engineering during the past thirty years: Commemorative volume for the thirtieth anniversary of the founding of The Society of Chinese Engineers.
2 vols., Zhongguo Gongchengshi Xuehui 中國工程師學會, Nanjing, 1946.
[There is also a repr., Wenhua Shuju, Taibei, 1967].

Zhou Kaiqing (*1972*) 周開慶.
Sichuan jingji zhi 四川經濟志.
The economy of Sichuan.
Shangwu Yinshuguan, Taibei, 1972.

Zhou Lisan 周立三, Hou Xuetao 侯學燾, & Chen Siqiao 陳泗橋 (*1946*).
Sichuan jingji ditu ji 四川經濟地圖集.
Economic atlas of Sichuan.
2 vols. (atlas + *shuoming* 説明), Zhongguo Dili Yanjiuso 中國地理研究所, Beipei 北碚, 1946.

Zhu Jieqin (*1985*) 朱杰勤.
古代的廣東（上中下）.
Ancient Guangdong (parts 1–3).
Guangzhou yanjiu 廣州研究 (Research on Guangzhou), 1985, no. 2, 59–65; no. 3, 60–63; no. 4, 72–73.

Zhu Peng (*1984*) 祝鵬.
Guangdong sheng Guangzhou shi Foshan diqu Shaoguan diqu yan'ge dili 廣東省廣州市佛山地區韶關地區沿革.
Historical geography of Guangzhou Municipality, Foshan Region, and Shaoguan Region, Guangdong.
Xuelin Chubanshe 學林出版社, Shanghai, 1984.

Zhu Sihuang (et al., eds.) (*1948*) 朱斯煌.
Minguo jingji shi 民國經濟史.
Symposium on the economic history of the Republic of China.
Yinhang Xuehui 銀行學會 & Yinhang Zhoubao She 銀行週報社, Shanghai, 1948.

Zhu Yulun (*1940*) 朱玉崙.
　　小規模煉鐵廠—解決後方鋼鐵需要之一途徑.
　　Small-scale ironworks – one means of filling the need for iron and steel in the
　　Rear.
　　Kuangye banyuekan 礦業半月刊 (Mining semimonthly, Chongqing), 1940,
　　3, no. 7/8/9/10, 1–3.

Index

Agricola 64, 65

Alley, Rewi 11, 15, 32, 34, 52, 78

alumina (Al$_2$O$_3$) 21, 23, 25, 33

Anhui 15, 17, 18, 27, 28

anthracite 48, 49, 50, 51, 53

archaeology 68

Beijing–Hankou Railway 15

Belgium 8

bellows 32, 33, 34, 49, 67

Bessemer converter 13, 45

Bibliothèque Nationale, Paris 58

Biringuccio, Vannoccio 65

blast furnace 10–14, 16–23, 27, 30–39, 43–45, 47, 48, 57, 58, 61, 63, 64, 65, 67, 68, 69, 70–76

Boluo (Guangdong) 66

Britain 4, 5, 6, 7, 27, 35, 37, 41, 42, 44, 46, 68, 69

Bronson, Bennet 16, 31, 51, 73

Budapest 31

calcination 21, 23, 31, 32, 33, 37, 38, 60, 61, 63, 64

capital 2, 4, 9, 10, 27, 45, 52, 66, 71, 77

capital-intensive technology 2, 9, 77

carbon (C) 13, 20, 21, 23, 24, 25, 31, 39, 41, 51

cast iron 13, 21, 24, 25, 26, 27, 30, 32, 37, 38, 39, 42, 49, 50, 51, 52, 63, 66, 67, 72, 73

Changzhi (Shanxi) 56

charcoal 17, 18, 21, 22, 23, 24, 26, 31, 32, 33, 36, 37, 39, 40, 42, 43, 56, 64, 65, 66, 69

Chengdu (Sichuan) 30, 31

Chiang Kai-shek 46

Chinese Industrial Cooperatives 11

Chinese Maritime Customs 7

coal 8, 22, 23, 30, 32, 36, 39, 41, 42, 43, 48, 51, 52, 53, 56, 65

competition 1, 2, 5, 8, 9, 28, 44, 53, 54, 77

copper (Cu) 27, 36

Cort, Henry 39, 41

Cremer, L. 29, 30, 31, 33, 34, 35, 36, 39, 44, 45

crucible smelting 13, 48–52, 55–57, 76

cupola furnace 33, 51, 71

Curwen, Charles 11, 72

Dabieshan 4, 10, 12, 14, 15–28, 30, 41, 58, 63, 64, 76, 77

Datangji (Luoding, Guangdong) 66, 68, 70

Dayang (Shanxi) 48, 52

Deng Liangqin 36

diatomite 18, 65

Dongan (Guangdong) 65

Du Fu 53

dumping 4

E Mida 74

economies of scale 13, 14, 52, 76

efficiency 2, 5, 6, 10, 11, 13, 14, 41, 42, 46, 64, 68, 77

embryonic capitalism 71

English East India Company 4, 5, 6

entrepreneurs 3, 9, 27, 46

Europe 4, 5, 6, 8, 28, 32, 42, 45, 48, 51

Fenjie (Luoding, Guangdong) 70

fining 13, 21, 24, 25, 27, 39, 41, 42, 67, 70, 71
flux 16, 21, 22, 23, 31, 33, 34, 35, 38, 44, 46, 64
Foshan (Guangdong) 58, 64, 66, 67, 70, 71, 73
France 5, 58
fuel 18, 20, 21, 23, 25, 26, 31, 33, 34, 35, 36, 37, 41, 42, 50, 52, 64, 71
Fuling (Sichuan) 39
Ganbazi (Nanchuan, Sichuan) 34
gangue 23, 31
Gansu 46
Gaoping (Shanxi) 50
Geerts, A. J. C. 8
Germany 13, 32, 45
ginseng 7
Great Leap Forward 2, 10, 11, 15, 19, 20, 30, 42, 47, 55, 56, 68, 75, 78, 79
Gu Zuyu 26
Guangdong 2, 4, 7, 8, 9, 12, 14, 27, 52, 58–75, 76, 77
Guangdong Provincial Museum 68
Guangdong xinyu 63, 64, 65–67, 69
Guangshui (Hubei) 26
Guangxi 77
Guangzhou (Guangdong) 5, 6, 7, 58, 63, 70, 73
guilds 4
Guo Yujing 15, 18
Han period 43, 56, 68, 69
Hankou (Hubei) 27
Hanyang (Hubei) 9
Hara, Zenshirō 15, 51
Hartwell, Robert 43
Hechuan (Sichuan) 47
Henan 10, 12, 15, 16, 17, 21, 24, 25, 26, 28
Hirth, Friedrich 71, 73
Höganäs, Sweden 55
Hollister-Short, Graham 63
Huang Jiguang Furnace 17, 19
Huang'an (Hubei) 26
Huangnipu (Yingjing, Sichuan) 32

Huangpi (Hubei) 26
Hubei 9, 12, 15, 17, 18, 19, 20, 26, 28
Hunan 30, 36, 53
Huoqiu (Anhui) 26
Huoshan (Anhui) 26, 27
imperialism 2, 4, 27, 71
Industrial Revolution 5
investment 2, 8, 9, 14, 27, 57, 66, 77, 78
ironsand 16, 17, 18, 21, 26, 40, 41
ironworks 5, 9, 10, 15, 16, 21, 28, 29, 31, 33, 36, 44, 46, 52, 55, 58, 63, 64, 66, 68, 70, 72, 73, 74, 78
Japan 8, 9, 13, 44, 46, 51, 78
Jiangbei (Sichuan) 36, 47
Jiayu (Hubei) 18
Jincheng (Shanxi) 54
Jinzhai (Anhui) 18
Korea 17
labour 9, 11, 24, 40, 42, 45, 53, 64, 68, 76, 77, 79
labour-intensive technology 2, 10, 64, 77
'laying down the dollar' 6
lead (Pb) 6, 60
Liljevalch, C. F. 5, 6, 73
lime (CaO) 21, 23, 25, 28, 33, 35, 56, 68
Limestone (CaCO$_3$) 21, 23, 31, 33, 34, 35, 38, 44, 48, 55, 56
limonite 33, 34
Liu Wenhui 46
Liu Xiang 46, 78
Liyinba (Nanchuan, Sichuan) 34
local gazetteers 15, 26, 28, 30, 31
Luo Mian 30, 31, 34, 35, 36, 37, 38, 39, 40, 41, 45, 46
Luo Yixing 9, 64, 71, 73, 74, 75
Luoding (Guangdong) 66, 68, 69, 70, 71
Luojing River 70
Luxia (Luoding, Guangdong) 68, 69, 72
Macheng (Hubei) 17, 19, 20, 26, 28
magnesia (MgO) 21, 33

Mai Wenyuan 70
Manchuria 7
manganese (Mn) 25, 33, 39
Mexican dollar 5, 6
mining 8, 9, 15, 30, 33, 48, 53, 64, 66, 69, 77
monopolies 5, 7, 43
Nanjing (Jiangsu) 46
Nantong Mining District (Sichuan) 34
Needham, Dorothy 46
Needham, Joseph 29, 46
needles 9, 53, 55
Netherlands 5, 73
Ningxia 46
Nyström, Erik Torsten 15, 16, 17, 18, 21, 25, 28, 48, 51, 55
opium 2, 6, 7, 9
ore 9, 12, 13, 16, 22, 23, 27, 30, 31, 32, 33, 34, 36, 37, 38, 48, 50, 51, 52, 53, 59, 60, 61, 63, 64, 65, 66, 68, 71, 72
Percy, John 21, 31, 33, 35, 41, 42, 45, 68, 69
phosphorus (P) 25, 38, 39, 50
pig iron 10, 18, 25, 33, 35, 37, 38, 39, 40, 41, 44, 50, 54, 55, 64, 70, 71, 73
Pingding (Shanxi) 54, 55, 56
prices 5, 6, 8, 9, 10, 27, 28, 40, 45, 53
progress 46, 76, 77
puddling 13, 36, 39, 40, 41, 42, 43, 44, 70, 71
Qijiang (Sichuan) 33, 34, 36, 39
Qin Empire 14, 58, 74
Qing period 4, 7, 8, 26, 27, 28, 29, 31, 44, 61, 68, 69, 70, 71, 72, 73, 74
Qishui (Hubei) 18
Qu Dajun 63, 64, 65, 67, 68, 69, 70, 71
Qu Minghong 67
quality 5, 10, 11, 27, 30, 33, 37, 51, 55, 56, 66, 71, 77, 78
railways 2, 15, 55
Read, Thomas T. 1, 51, 55, 56
Red Basin (Sichuan) 29, 30
refractories 17, 18, 24, 34, 35, 36, 39, 41, 49, 51, 52, 63, 69

Richthofen, Ferdinand von 8, 29, 30, 36, 39, 48, 50, 51, 52, 53, 54, 55
Rongjing (Sichuan) 32
Rostoker, William 16, 31
Rowe, William T. 27
Russia 69
salt 7, 30, 36, 65, 69, 72, 74
sandstone 17, 18, 33, 36, 37, 39, 41
scissors 53, 55
scrap iron 7, 70, 71, 73
Shaanxi 29, 46, 78
Shangcheng (Henan) 18, 24, 28
Shanxi 4, 8, 9, 12, 13, 24, 39, 48–57, 76, 77
Sheffield 45
Shijiazhuang (Hebei) 55, 56
Shiwan (Boluo, Guangdong) 66, 67
Shuangshui River (Guangdong) 70
Sichuan 4, 9, 12, 14, 21, 22, 29–47, 76, 77, 78
siderite 31
silica (SiO$_2$) 16, 21, 23, 25, 33, 38, 41
silicon (Si) 25, 26, 38, 39, 50
Sino–Japanese War 2
slag 18, 21, 23, 24, 25, 31, 32, 33, 34, 35, 36, 37, 38, 40, 41, 44, 51, 61, 63, 64, 69
small-scale industry 2, 11, 14, 15, 27, 28, 58, 63, 64, 73, 77, 78
Song period 43
Southeast Asia 58, 73
steam 45
steamships 2, 44, 77
steel 2, 5, 13, 26, 27, 32, 36, 45, 46, 50, 54, 67, 73, 78, 79
Sui (Hubei) 26
sulphur (S) 23, 25, 26, 31, 33, 38, 39, 42, 44, 50, 55, 56, 77
Susong (Anhui) 26, 28
Széchenyi, Béla 29, 30, 31, 32, 37, 45
Taiping Rebellion 77
Taiyuan (Shanxi) 55, 56
Tang period 15, 25, 53
taphole 17, 18, 23, 34, 36, 37, 66, 69
tea 2, 6, 9

Tegengren, F. R. 7, 9, 15, 16, 17, 18, 21, 25, 28, 29, 30, 31, 32, 51, 52, 54, 55, 56

temperature 17, 20, 22, 23, 24, 37, 39, 40, 41, 42, 55, 68

transportation 5, 9, 10, 11, 14, 15, 28, 30, 44, 45, 53, 55, 64, 70, 71, 76, 77, 78

Treaty of Nanjing 7, 8

treaty ports 8

tuyère 22, 23, 33, 34, 36, 37, 69

Wang Zhuquan 54

Wanshengchang (Nantong, Sichuan) 34

warlords 46, 78

water power 30, 31, 32, 36, 44, 45, 68

watercolours 58

Weiyuan (Sichuan) 33, 39, 45

windbox (double-acting piston bellows) 18, 25, 36, 63, 68

wok 26, 27, 51, 66, 70, 71

wood 23, 24, 34, 36, 42, 64, 65, 66, 68

World War I 2, 10, 28, 54, 56, 77, 79

World War II 2, 30, 31, 46

wrought iron 7, 8, 12, 13, 24, 25, 27, 39, 40, 42, 49, 50, 51, 52, 55, 67, 70, 72, 73

Wuhu (Anhui) 27

wustite (FeO) 21, 23, 25, 31, 33, 41

Xiadian (Yingshan, Hubei) 26

Xijiang ([West River], Guangdong) 70

Xinyang (Henan) 15, 16, 17, 18, 21, 24, 26, 28

Xishuanghe (Xinyang, Henan) 27

Yang Kuan 24, 30, 39, 42, 43, 51, 54, 55, 57, 68

Yangzi Gorges 29, 44

Yangzi River 29, 30, 44, 77, 78

Ye Qihua 70

Yincheng (Shanxi) 48

Yingjing (Sichuan) 32

Yingshan (Hubei) 26, 28

Yunfu (Guangdong) 65, 71

Yunnan 30

Yutai 27

Zengcheng (Guangdong) 67

Zezhou (Shanxi) 54

Zhan Ruoshui 67

Zhang Zhidong 75

Zhaoqing (Guangdong) 70

Zhou Enlai 10

The Nordic Institute of Asian Studies (NIAS) is
funded by the governments of Denmark, Finland,
Iceland, Norway and Sweden via the Nordic Council
of Ministers, and works to encourage and support
Asian studies in the Nordic countries. In so doing,
NIAS has published well in excess of one hundred
books in the last three decades, most of them in co-
operation with Curzon Press.

Nordic Council of Ministers